A Place to Live and Work

A Place to Live and Work

The Henry Disston Saw Works and the Tacony Community of Philadelphia

Harry C. Silcox

THE PENNSYLVANIA STATE UNIVERSITY PRESS
University Park, Pennsylvania

Library of Congress Cataloging-in-Publication Data

Silcox, Harry C., 1933–
A place to live and work : the Henry Disston saw works and the Tacony community of Philadelphia / Harry C. Silcox.
p. cm.
Includes bibliographical references and index.
ISBN 0-2710-3075-5
1. Henry Disston & Sons, Inc. (Philadelphia, Pa.)—History.
2. Saw industry—Pennsylvania—Philadelphia—History. 3. Tacony (Philadelphia, Pa.) I. Title.
HD9529.S294H467 1994
338.7'62193—dc20 93–9923
CIP

Printed in the United States of America

Published by The Pennsylvania State University Press,
Barbara Building, Suite C, University Park, PA 16802-1003

It is the policy of The Pennsylvania State University Press to use acid-free paper for the first printing of all clothbound books. Publications on uncoated stock satisfy the minimum requirements of American National Standard for Information Sciences—Permanence of Paper for Printed Library Materials, ANSI Z39.48–1984.

To

LOUIS A. IATAROLA,

a man who has taken pride in Tacony and its people, as shown by his restoration of the Music Hall. His efforts have created a new spirit in Tacony that brings back memories of the town as it was in the days of Henry Disston.

Contents

List of Illustrations

Figures 1–27 follow page 44.

Figures 28–55 follow page 108.

List of Tables

Preface

I am a Taconyite, born some sixty years ago in the front bedroom of 6606 Torresdale Avenue. This was not unusual for Tacony residents in those days, since there was no hospital in the town. My earliest childhood memories were of the streets of the town, Boy Scout Troop 24 (then one of the oldest in the country), and the sports teams that represented the community. I never quite understood why the long streams of workers headed for the river each day, or why there were no saloons anywhere. I played baseball on one of the city's finest fields—Disston Ballpark, at Unruh and State Road. I attended the Mary and Hamilton Disston grade schools and played at the Disston Playground every day, without knowing who or what the name represented. I would never have known had I not become interested in the history of Philadelphia and found that the Disston name kept appearing in the documents I read.

When I was in the midst of researching a book on the Irish-American politician William McMullen, my attention was attracted to his friend, Hamilton Disston. Both were volunteer firemen who shared a love for politics. "Ham," as he was known to his friends, was the son of the wealthy saw manufacturer Henry Disston, who had established a utopian industrial community in Tacony, the same community in which I was born. I was fascinated to read accounts of how he kept this little community under his control, allowing no competing factories and arranging for well-ventilated homes for his workers in the residential area. It was then that I began gathering research materials about Henry Disston & Sons and Tacony.

An 1887 document intrigued me more than the others. In it, Pennsylvania's secretary of the interior lauded Henry Disston's achievement of the perfect relationship between worker and factory owner. The secretary urged Pennsylvanians to go to Tacony and see the ideal industrial town, in the belief that the labor unrest endemic in the nation during the 1880s would fade if factory owners adopted Disston's methods.

It soon became clear that Tacony was not the usual urban industrial community of its day. It was a planned urban community in the suburban countryside, a small, one-factory town controlled by a single family for over eighty years. Today, the community still contains most of its original 1880s factory buildings and homes.

At first I was fascinated by the way the Disstons maintained control over their industrial town of Tacony—known as paternalism. This theme dominated the first article I wrote on the subject for the *Pennsylvania Magazine of History and Biography*, "Henry Disston's Model Industrial Community: Nineteenth-Century Paternalism in Tacony, Philadelphia." But as I began to enlarge my study and trace the effects of the Disstons' own brand of paternalism on the town and the people, I became aware that the rapid social, economic, and political changes of the twentieth century had overwhelmed paternalism as a force in the community.

It was then that I realized that the dynamics of events at Disston during its 150-year history were similar to those of most family-owned firms in America. Most such firms had their birth in antebellum America, most practiced some form of paternalism, most reached an economic zenith by the mid-1920s, and most had disappeared by the mid-1950s, swallowed up by corporations that were more able to apply new national financing and multiproduct ownership methods to locally minded and one-product firms.

The important factors in the history of these firms were family ownership, the way profits were accumulated, new-product development, the use of industrial welfare, emerging unionism, and the effects of the two world wars on production and on firm survival. The years before and between these wars also became a vital part of the story, as did the Great Depression. The sale of Henry Disston & Sons in 1955 embodies the story of an acquisition man who purchased the company to sell off parts of it for profit and of his effect on the community. All these issues provide a platform for analysis that has more far-reaching implications for our society than the issue of paternalism in a bygone day.

The book is divided into two basic time periods. Chapters 1 through 4 cover years up to about World War I, during which period paternalism was an important element in the relationship between company and town. In these chapters I review the life of Henry Disston, the reasons for his financial success, and the effects of this success on the people of Tacony. Chapters 5 through 9 examine the methods the company used in the twentieth century to adjust to changes in society and still maintain some control over the labor force and the community. National and world events had an impact on this relationship: World War I, the period of internal-products development in the 1920s and 1930s, the Great Depression, unionism, World War II, and the selling of the company in

1955. Interspersed throughout each of these chapters is the story of the people of Tacony.

The Tacony area itself encompasses land in Northeast Philadelphia located west of the Delaware River between Levick Street and Cottman Avenue, a block short of Frankford Avenue. In all, it covers about 500 acres. Much of the area is called Mayfair today, but the original planned utopian community remains relatively untouched by the encroachments of modern society. As such, it has much to offer the industrial and social historian.

Philadelphia has been described as a city of neighborhoods. This definition has misled the public into seeing neighborhood boundaries as rigid and static. Ward, parish, police district, and postal zip code boundaries are precise but rarely the same. In reality, neighborhoods in Philadelphia are much like amoebas, rearranging their shape and form depending on the way they are viewed or nurtured, or the time frame in which they are studied. In some instances, a portion of an old neighborhood will break off and form a new neighborhood, or one neighborhood will expand and encompass others. Real-estate values and developers also influence the size and shape of neighborhoods.

Transportation systems have forced many changes in neighborhoods throughout Philadelphia's history. The increased speed of transportation lines (from carriages to trolleys to high-speed lines that increased the radius of home base while keeping workers within twenty-five minutes of work) made a greater choice of living areas available to workers. "Main streets," originally laid out perpendicular to the Delaware River, were quickly superseded by the street that had a trolley line. Trolley lines emanated like wheel spokes from the center of Philadelphia (Center City), with few public connections between adjacent towns of the city's northeast section. All trolleys led to Center City. Movie houses and stores followed the transportation lines in the 1920s and 1930s, creating many small shopping villages throughout the city. Clearly, the study of neighborhoods over time is tied to changing transportation modes, and this is especially true of Tacony (see Chapter 4). For most of its history, the Disston company drew its workforce from a specific community, and the members of that community looked to the Disston company and family for employment.

The story that unfolds here is one of a factory and a community bound together by a culture that was dominated by the industrial revolution. The people of Tacony remained loyal to the Disston company, but this loyalty proved ultimately to be detrimental to them. The eventual sale of Disston & Sons forced negative economic changes on Tacony, a story repeated in town after town in post–World War II America. Trusting their future to H. K. Porter President Samuel Mellon Evans, the new owner of Disston & Sons, proved disastrous to the community. Elsewhere in America, community after community felt the same sting of abandonment as companies took their business elsewhere in a never-ending search for profits.

Another valuable insight from the Tacony/Disston story is the importance of work to the citizen. Work can be seen as a source of energy building up and supporting a community. It helps to transform a pristine landscape into an urban village. It determines migration, family structure, social connections or disruptions, labor peace or unrest, and social arrangements within a community. In a small community like Tacony, recreation, entertainment, and social activities were related as much to work as to churches, schools, and family. To appreciate the nature and interaction of the nineteenth-century community, social historians need to spend more time examining work opportunities as a major force sustaining or challenging urban institutions.

The interaction of neighborhood and factory once animated Tacony. Tacony became a factory community because the proprietary capitalist Henry Disston encouraged skilled laborers to make a lifelong commitment to his company. Homes and families around the factory made for worker stability in an age when transportation to rural worksites was limited. Steady work was what nineteenth-century employees sought most. Disston's skilled workers carried this priority into the twentieth century. Henry Disston too believed that work strengthened family life, community prosperity, and public and private orderliness. His experiment showed that when industrialists cared about their workers, the neighborhood flourished.

Acknowledgments

One might legitimately ask, "Why write a book about a single industry in one city?" The answer is simple: Few if any histories of family firms span a century and a half and examine the profound effect of a factory on the life-style of a local community. The firm of Henry Disston & Sons is particularly worthy of study because of the richness of the materials available, the stability of the community, and the information the company's story provides about the history of family-held firms in general.

I was encouraged to pursue this study by Phil Scranton, who allowed me to be part of a Philadelphia Alliance for Teaching Humanities in School summer workshop on Industrial Philadelphia in 1988 so that I might begin some serious research on the subject. He also read a rough draft and suggested a stronger focus on the Disston company than I had originally planned. His persuasiveness and confidence in my ability were the primary force behind my being able to complete the book. Also important was the sharp but needed criticism of Walter Licht of the University of Pennsylvania, who suggested the ultimate chapter arrangement for the book.

Expert research help was given by Tina Lamb of the Tacony Public Library and Fred Miller at Temple University's Urban Archives. Special thanks go to John Alviti, executive director of the Atwater Kent Museum, for his insights into the study of neighborhoods; to Randall Miller, editor of the *Pennsylvania Magazine of History and Biography;* and to Bob Eskin, curator of the photographic collection at the Atwater Kent Museum for their permission to use materials in this work. John R. Bowie, of the Oliver Evans Press, was most helpful in identifying new directions for the work. His editing and publication of the book *Workshop of the World* helped me gain much needed background about other manufacturing firms in Philadelphia. Research done at the Hagley Museum in Delaware, the Pennsylvania Historical Society, and the Newspaper Department of the Philadelphia Free Library at Logan Square was facilitated by cooperative and competent staffs. And finally, everyone in Tacony who stopped me on the street to tell me a story about the town is as much a part of this book as the people whose names I use, for it is the town as a whole that reflects what Henry Disston was trying to do. Everyone in Tacony can find himself or herself somewhere in this book.

Most important to the study was William Leeds Disston, who knows more than any living person about the meetings and legal actions that surrounded the sale of the company to H. K. Porter. His reading and correction of the manuscript helped clarify much of what took place during the 1940s and 1950s. His written recollections of the Disston company will stand as a reference for anyone studying the history of industry in Philadelphia. I am grateful to William L. Disston for his support and friendship throughout this study, and to his son, Morris Disston, for his cooperation.

Henry Disston, grandson of Jacob Disston, allowed me use of the family history and permitted me to copy many of the photographs in this book. His honesty and frankness were a comfort to me as I researched the family split over the sale of the company.

Crucial to the completion of this manuscript was Barbara Horwitz, who revised and edited the first rough draft. Her friendship and patience enabled me to overcome early frustrations concerning the work. Jane Barry, with her usual skill, focused each chapter's theme, revising the order of events in the original manuscript. Her advice is always thoughtful, her editing clarifying, and her patience that of a saint. Emma Kapiliovich and Alice Parsha were instrumental in having the final computer manuscript formatted for printing. Finally, author Bruce Stutz aided in the research, uncovering much about Samuel Mellon Evans's life.

Documentary sources include surviving records of the original firm found at the final Disston & Sons factory site at Longshore and State Road in Tacony. Longtime employee Roland Woehr helped in providing access to records and in reconstructing sequences of events and the lives of those who worked in the factory. Parish histories of St. Vincent's Home and St. Leo's Church were helpful in understanding early events in Tacony. A bound copy of the *Disston "Bits"* (the company's monthly paper), in the possession of William Rowen, and the copies of the *Disston Crucible* (the technical magazine for the lumbermen), at the Logan Square library, were also of great help. Discovery of three bound volumes of *Philadelphia Made Hardware* at the Disston plant clarified the importance of the hardware business to Henry Disston & Sons.

The people of Tacony—Disston employees, their relatives, and the community residents alike—were most helpful: William Rowen, master smither; Bob Bachman, Niles-Sheppard crane operator; Roland Woehr, engineer at Disston Precision Company; Mrs. Eleanor Shuman Dick, daughter of inventor Frank Shuman; Joseph Rosenwald, owner of the Foodarama grocery store; Edward Darreff, Darreff's grocery store; Charles Norbeck, Disston worker; Fred, Arnold, and Ernest Rodgers, of Jewitt & Rodgers, great-grandsons of Jonathan Marsden; Russell McIntyre, Madison grinding machine operator and union secretary/treasurer; Allen H. Wetter Jr., a descendant of the Whittaker family; Judith Zaslofsky, Hyman Rubin's daughter; Mark Ward and Ron Turfitt, current employees of Disston Precision; Catherine Seed, secretary of the Mary Disston Estate; Naomi Vickers, daughter of Billy Seed; John Hansbury, former Disston worker; Marguerite Dorsey Farley, sister of Tacony Congressman Frank Dorsey; and Thelma Koons, Disston office secretary, 1924–1980.

The inspiration for this work came from Louis Iatarola, owner of the Tacony Music Hall. Against all odds, Iatarola decided to restore a 105-year-old building located in the least attractive part of old Tacony to its original splendor. The building stands as proof of Tacony's better days. His faith in the community and his continuing efforts to improve the neighborhood have set an example for those around him.

Finally, to my wife, Shirley McKay Silcox, goes a special thanks for tolerating the hours that the computer disturbed the peace of our household. She alone knows the sacrifices required in the cause of research and writing. Together we have learned more than we want to know about computer crashes and the frustrations of the new technology. The writer's only consolation is that others will read the work and gain insights from it.

CHAPTER

1

Henry Disston Founds a Business and a Community, 1819–1878

IF EVER AN INSTITUTION was the shadow of a man, it was the Disston company. To know the man is to understand the Disston company's methods and purpose. Henry Disston, founder of what was originally named the Keystone Saw, Tool, Steel & File Works, was born in Tewkesbury, England, on May 24, 1819. His family had been textile-mill owners and manufacturers. His grandfather William Disston owned Nafford Mills near Tewkesbury, where Henry's father, Thomas, began work as a young boy. When Henry was four years old, the family moved to Derby, in Nottingham, where Thomas, now a machinist, found a job manufacturing lace machines. Nine years later Thomas invented a new machine for making a special fine lace. Aware that no product of comparable quality existed at this time in America, a group of British businessmen offered to pay Thomas's transportation and share the potential profits if he would bring the machine to a mill in Albany, New York. Thomas accepted the offer. Lacking the money to take his entire family with him to America, he decided to bring his daughter, Marianna, to keep house, and his eldest son, Henry, to act as a machinist apprentice. The trip ended tragically when Thomas died three days after arriving in Philadelphia in 1833. The lace machine was taken from the ship and sold to an unknown buyer, and the money was sent to his widow in England. The two children were destined to remain in America.[1]

Friends quickly found Marianna work in a private home, but finding the young Henry a place under a machinist proved more difficult. Eventually, he apprenticed himself for seven years to Lindley, Johnson &

Whitecraft, a Philadelphia saw-making firm founded by three Englishmen. There he learned about saws and also began experimenting, like his father, with craft machines. By the age of twenty-one he had accumulated $350 in savings, enough to start his own saw business in the spring of 1840. Initially, he did all the work himself, but in July he hired his first apprentice, David D. Bickley, and carried on the business in the back room of a building at Second and Arch Streets. Soon he moved to Broad Street, opposite Letter Lane, where Disston produced his first saw. He wheeled coal from the Willow Street wharf to the basement of his house, which he had converted into a hardening shop for making saw blades. The blades were then wheeled to a grinder located at Second Street and Kensington Avenue. Handles were then purchased and the saw was assembled for sale. Working a full day, Disston and Bickley were able to produce about a dozen saws a day.[2]

For the first three years business was slow. At the time, Americans preferred foreign-made saws, usually those from Sheffield. Disston found himself making saws three days a week and spending the next three trying to sell them. On Saturday night he delivered orders to the stores, and on Sunday he went to church. In 1844 he rented space and borrowed $200 to outfit a shop with the first steam-powered saw machinery in America. Unfortunately for Disston, he had been bilked. The lessor had himself leased the building and was behind in the rent. The real owner summoned the sheriff and seized all Disston's equipment for back rent.

Disston still had a few unfinished tools and saws in his possession. He finished them in his home to earn the money to buy back his equipment. His new landlord doubled the rent, prompting Disston to move to Third and Arch Streets. Two years later this landlord too forced him to move, and Disston rented a frame building at Front and Laurel Streets in Kensington from William Mills. A fire destroyed the building in 1849. By now, Disston was thoroughly disillusioned with the idea of renting, and so he purchased land next to the fire site and built his first factory.[3]

Early on, Disston earned a reputation for quality that distinguished him from his Philadelphia counterparts, and he used this reputation to gain the trade of individual hardware men in the neighborhoods of the city. A story told of Disston during this early period indicates a willingness to sacrifice profit for quality. Disston personally delivered an order of saws to a hardware store. As he placed the order on the counter, he noticed that one of the blades was soft. Immediately, he had the entire order removed from the store. The merchant asked him to stop, since he needed saws to sell and some of the saws might be usable. Disston refused, explaining that all the saws had been tempered together and that he would not risk having inferior saws bear his name. Stories like this—

whether true or not—served to make the name Disston synonymous with integrity and quality.[4]

The making of a saw at that time required steel ingots, which were lightly hammered and then rolled into sheets of varying thickness. Saws were cut from this stock. The teeth were hand punched, and holes were drilled for the handle. The steel was the crucial element in the process. Americans had difficulty making steel hard enough to bear up under the heat and pressures of sawing. Most U.S. saw-makers at the time imported steel from Sheffield, where the English had mastered the technique of using fired clay pots (or "crucibles") to produce high-quality steel. In 1855, Disston became the first American saw manufacturer to bring steelworkers from Sheffield and open a crucible mill. This experiment with crucible steel did not mean that Disston discontinued the use of English steel. On the contrary, an English steel agent noted in 1867 that although Disston made steel for his large saws, he bought steel for circular saws from Jessop's in Sheffield. Disston was still using Sheffield steel in some operations in the 1880s.[5]

In 1859 Disston had 150 men working for him, more than any of his local competitors: Walter Cresson's Saw Works of Conshohocken; William Rowland's Saw Works in Cheltenham, which produced the first American-made saw in 1802; and William Conaway's Saw, of 402 Cherry Street. To gain an edge over these competitors, Disston used his family connections in England to keep him abreast of trends in steel-making. His machinist brothers were able to acquire the use of John Sylvester's patented process for tempering and restoring the shape of scraps of hardened steel. Sylvester had invented a large press that used a turn screw to compress heated steel scraps into ingots. Disston had the device brought to America so he could reuse the steel cuttings from his saws. This cost-saving technique gave him an advantage over other saw manufacturers, who were still sending their scrap metal back to England to be remelted. His recycling method yielded steel faster and cheaper.[6]

Not to be outdone, Disston's competitor Walter Cresson began using Sylvester's process, apparently without the inventor's permission. In 1857 Disston brought suit for patent infringement on Sylvester's behalf before the Circuit Court of the United States. Sylvester remained in England throughout the two-year court battle, which Disston aggressively pursued. Cresson's defense was that the process had been used extensively in England prior to Sylvester's patent from the Crown. Therefore, he argued, the process was in the "public domain" and could be used by anyone. Ultimately, the judge ruled against Sylvester, but Disston learned from this incident that the best method of protecting new devices and processes was to keep them secret.

Despite Disston's inability to protect Sylvester's patent, few manufacturers could match his line of saws, either in quality of manufacture or in quantity of styles. His first catalog in 1855 listed twenty-one different saws, four types of knives, trowels, gauges, saw cutters, all kinds of springs, steel-blade squares, and bevels. Six years later Disston's company, Keystone Saw Works, was described by the *Philadelphia Press* as the biggest saw factory in the country.[7]

Throughout the 1860s Disston found himself with back orders because he did not have enough skilled workers. There was always work for newly arrived skilled craftsmen from other U.S. cities or from the Continent and England. One example of how fast an experienced craftsman could get work was recounted in the *Disston Crucible* by Fred Smith, an apprentice at a saw works in Albany, New York, when it was closed because of a lack of orders. After signing some papers, Smith was given his freedom and set out the next day for Philadelphia. On Monday morning, March 10, 1861, he arrived at the old Kensington railway station and asked the way to the Keystone Saw Works. As he walked toward the factory, a chance meeting with David Brickley, Disston's contract maker for all long saws, got him the job. Smith rented a room and, though tired from the long trip, was at work by twelve noon the same day. Henry Disston stopped by within the hour to welcome him to the firm and nicknamed him "my runaway apprentice."[8]

Despite Disston's remarkable growth, he was not immune to turns in the economy. Like other northern businessmen, he experienced the effects of the 1857 panic and the ensuing depression. The secession crisis in the latter part of 1860 deepened the business turndown. As the southern states seceded from the Union, their businessmen refused to pay bills for goods already received from the North. Diarist Sydney George Fisher lamented that the Market Street store owners were unable to pay their bills because southern businessmen had defaulted on their debts. A credit reporter at the time, R. G. Dunn, estimated in 1862 that unpaid southern liabilities had cost northern businessmen nearly $300 million.[9]

For Henry Disston, however, the Civil War became the means by which his company would double in size and be positioned at the war's end to capture a sizable share of the national saw market.[10] By 1862, Keystone Saw Works had switched some of its operation into war supplies. Sabers, bayonets, knapsack mountings, and guns gave Henry Disston diverse orders to fill beyond the continuing demand for saws. The Morrill tariff of 1861 added to Disston's profit margin by placing import taxes on iron entering the country. Saw-makers who had not built their own steel mills were virtually put out of business by this act.[11]

Another war-related source of business came from Philadelphia shipyard owner William H. Cramp. Cramp had recently changed his thirty-year-old business from wood to iron vessels, a fortuitous move that put him at the forefront of technological innovation and guaranteed him an oversupply of government war contracts. By 1863 these contracts had increased his need for steel plate both for new ships and for conversion of wooden blockage vessels. Keystone Saw Works became one of Cramp's major suppliers of steel.[12] Soon after, Disston built a rolling mill to accommodate Cramp's needs. As orders increased, so did Disston's profits, giving him the capital he needed to expand.[13]

The key to Disston's growth remained his passionate search for new machinery that would increase production without diminishing the quality of the product. He encouraged everyone at the firm to think of ways to improve the machines they used and the products they sold. Saws were manufactured in every conceivable way. In the late 1860s, Disston produced inserted-tooth saws; later, cross-cut saws were designed with raked teeth, making them by far the most effective saws used in American lumber camps.[14]

Disston's visit to Paris in 1865 resulted in a marked advance in the making of saw handles. Hearing of a superior bandsaw that he might use to cut handles, he visited the factory and purchased two of them. The key to the operation was the speed of the revolving saw. At first, the men in Disston's factory were fearful of using it lest the three-eighths-inch steel blade break and injure them. To show that the new machine could be used safely, Hamilton, Henry's eldest son, operated the first bandsaw set up in the factory. Disston prevailed, and the new saw dramatically increased the production of saw handles, from 20 dozen a day made by one man to 165 dozen a day made by two.[15]

Few available records refer directly to employment figures at Disston's saw works during these Civil War years, but the growth of the Disston company can be examined through the contemporary writings of Philadelphia manufacturing expert Lorin Blodget. By using employment figures for Philadelphia workers between 1860 and 1870, we can make some inferences about the increased number of employees at Disston. In 1860 the city had 562 steelworkers; by 1870 the number had risen to 1885. This 235 percent change made steel the fourth-largest employer among Philadelphia industries. Another business observer of the time, Edwin T. Freedley, analyzed the change in total output of products in the city between 1860 and 1866. His findings give credence to the claim that the Disston company increased its saw business dramatically during the war. Sales of saws increased by 251 percent, from $271,000 in 1860 to $950,000 in 1866. Even when adjustments are made for inflation, the growth of the saw business was phenomenal. Because Disston was the major saw-maker in Philadelphia, most of the growth was his.[16]

Another key to understanding the company's growth is the Disston family. Henry took responsibility for all his brothers and sisters and gave them the opportunity to come to America. In the end they provided leadership within the firm that he could trust and that he needed to expand his saw works. The togetherness of the Disston clan continued into future generations, and the family name came to be identified with the factory more than its original name, the Keystone Saw Works.[17]

By the time of the Civil War, Henry and Mary Steelman Disston had nine children (not all lived past childhood) and needed a larger house. In 1864 Disston purchased a lot at 1515 Broad Street, about three miles from the factory, and built his first house. In 1870 the house was razed and replaced in 1872 by an elegant mansion constructed at the same location, providing ample evidence of the family's wealth and Henry Disston's position as one of Philadelphia's leading industrialists.[18]

The death of Henry's father had forced his second-oldest brother, William (1820–1872), to remain in England to care for his mother and younger brothers and sister. But brothers Charles (1823–1898) and Thomas (1832–1897) and sister Susan (1826–1898) came to America in 1846, and the brothers immediately went to work in the saw factory. Back in England, William was ready to give Henry whatever information he needed about steel processes or the new machinery then appearing on the European market. It was not until May 1868 that William migrated with his family, which included his son Henry (1845–1920), with the express purpose of founding a jobbing shop at Keystone Saw Works. The shop was organized to fill special requests for tools or other steel products and functioned very much like an experimental or developmental laboratory. The value of the shop to the firm is demonstrated by the listing of Thomas Disston, longtime subpartner in the shop, on twenty-seven of the company's forty-four patents.[19]

The English connection continued when William died in 1872. His son Henry returned to England two years later and remained there for twenty years as a salesman of tools and recruiter of personnel for the Disston factory.

Sales in England and recruitment of workers alone were not the only factors in the company's expansion, however. Important to Disston's growth was its willingness to produce new and better products for Americans. The story of Disston's discovery of the skewed-back saw typifies his inventive style. Early one morning in 1873, the story goes, Disston sought out plant superintendent Albert Butterworth. "Al," he said, "I'm not satisfied. There must be some other way to improve the handsaw even more. Why, last night I was going over the history of Egypt and Rome, and from some of the illustrations the shape of the saw blade today is about the same as then. I've been thinking about it all night. Get a piece of chalk. Now, draw a handsaw blade down there." Butterworth drew the blade on the floor, and Disston recognized instantly what he wanted to change. "See? There's more blade there than is required. It's too wide. . . . Just cut off a section of the back . . . curve it." Butterworth did as he was told, and the company had a new product that took less steel to make and functioned better for sawyers because it was light and had less blade friction.[20]

New-product success prompted the need for more production space. This, combined with the accumulation of capital during the Civil War years, stimulated discussions about moving the factory to a new site. After considering a number of locations, Disston settled on land in a small hamlet in the northeast section of Philadelphia called Tacony. This area, described as early as 1679 as a "village of Swedes and Finns," is located where the Pennypack Creek feeds into the Delaware River. Tacony came into being during the early part of the nineteenth century with the building of the Buttermilk Tavern on the Delaware. Tacony in this period was a summer vacation spot for Philadelphians and farmers of the area. The initial attempt to connect Tacony with the city occurred in 1846, when William H. Gatzmer secured a charter for the Philadelphia & Trenton Railroad. The people of Kensington refused to allow the railroad into the city, so Tacony became the terminus. Passengers going to the city left the train at the Delaware River and what is now Disston Street and took a boat to the Walnut Street wharf. A railroad hotel was built to accommodate the passengers, and a small community composed of steamboat and railroad workers quickly took shape.[21]

Henry Disston first visited Tacony when his brother Thomas purchased a number of lots from the Tacony Cottage Association in July 1855. This association was a private group that had arranged a speculative land deal to support the building of St. Vincent's Catholic German Orphanage, purchasing two farms totaling forty-nine acres at a cost of $19,000 and divided half the land into vacation cottage lots. Sale of these lots allowed the association to recover the purchase price and build the orphanage on the remaining land. Thomas built a summer home on what was described as one of the most beautiful and healthful spots along the Delaware. Easily accessible by steamboat or train, the Disstons used the area for summer vacations until Henry's wife had a summer home built in Atlantic City. Fish were plentiful in that area of the Delaware, and the nearby transportation made it a short trip to the city. Eventually, Henry Disston purchased a six-acre lot that included the Buttermilk Tavern.[22]

Not only was land cheap, but the location had two major sources of transportation. The Delaware River, with a dock already in place, was ideal for movements of coal and iron ingots. The railroad provided a means of transportation to the West, where saws were in continuous demand.

One element was still needed before Disston could consider moving his factory to Tacony: a supply of skilled labor. This sleepy farming-fishing village, home of a few railroad workers, some watermen, and random steamship employees, could not satisfy Disston's labor requirement. It would have been folly in 1872 to attempt to move the nearly 900 employees from the Front and Laurel Street factory to Tacony. There were simply not enough houses in the area, and the two hotels could board no more than thirty men. This meant that migration to Tacony would have to take place in stages over a number of years. First, Disston opened a small mill on the property as a laboratory to test experimental saws. Sawyers from the Laurel Street plant moved into a nearby hotel, and the experimental station began testing saws from the Laurel Street factory. The migration took twenty-seven years (see Table 1), with but a small portion accomplished during Henry Disston's lifetime. His vision of a residential area for his workers would be completed by his wife, Mary, and his children.[23]

This gradual transition allowed the transfer to take place without disrupting business. Except for the initial move of the jobbing shop, the

Table 1. Chronology of Transfer of the Keystone Works to Tacony

Ground broken for mill at Tacony	1872
Moved Handle Shop	1872
Moved File Shop	1873
Began building Steel Works	1877
Moved steelworkers	1879
Moved Long Saw Dept.	1881
Moved Hardening Dept.	1881
Moved Circular Saw Dept.	1882
Moved Jobbing Shop	1883
Moved Handsaw Dept.	1884
Jobbing Shop moved back to Laurel St. because of inconvenience to customers	1885
Moved Square and Level Dept.	1887
Moved Butcher Saw and Trowel Dept.	1896
Moved Jobbing Shop	1899

SOURCE: Paul N. Morgan, "The Henry Disston Family Enterprise," *Chronicle of the Early American Industries Association* 38 (June 1985), 19.

plan to open new factories in Tacony went as scheduled. Not only could Disston expect to increase his profits from an enlarged factory, he stood to gain from real-estate development as well. Cheap farmland offered great potential for profit, even if money had to be spent on community needs such as streets and a school. Development of this farm area had two phases: attracting skilled workers and building houses for them.[24]

William Smith, a cousin of Henry Disston and later the firm's master mechanic, remembered well the building of the first Disston factory in Tacony. On the last Thursday night in September 1872, Henry Disston, Smith, and Samuel Bevan (then master mechanic), were in Tacony marking out the corners for the building. They were discussing the talk in Philadelphia that Disston was moving "too far out of town for his own good." Disston insisted that they start that night to dig the foundation: "Tomorrow's Friday. I'm not a bit superstitious, but those fellows who advised us not to move up here would certainly have a good laugh if anything went wrong, and some would surely say it was because we started on Friday." That night the three men dug the foundation for half the side of a building. It was to become the file works.

In November 1872 fire destroyed some of the Laurel Street plant, forcing Disston to set up a temporary handle shop in the new building in Tacony. The relocation of the entire file works would have to wait until 1873. The most difficult operation to move was the steel plant. So it would be done right, master English smelter Jonathan Marsden moved to Tacony in 1875 to plan and supervise the construction of the steel plant. The steelworkers followed four years later.[25]

In 1876 Henry Disston began work on his planned residential community, which lay to the west of the Pennsylvania Railroad's newly acquired and now directly connected Philadelphia to New York railroad tracks. Disston refused to use water from the Delaware River, which the rest of the city utilized. To ensure a pure supply of drinking water, he built a water-pumping station at Sandy Ford Springs, a feeder stream of the Pennypack Creek some two miles from Tacony. Water tanks at what is now Cottage and Disston Streets stored the water and provided a gravity feed to the pipes laid down Longshore Street to the factory. The community could gain access to the water from the Longshore Street line and pay the Disston-controlled Tacony Water Company for the service. This gave Disston's town the purest water in Philadelphia. (River water was still used in the steelmaking process.)

Houses for the workers followed. Eventually, the Disston family would own more than 600 homes, holding some 360 in the Mary Disston Estate until 1943. All other homes in Tacony were built by Disston and sold to the workers, or the land was sold outright by the Disston family. Homes could be rented or purchased, according to the worker's desire.[26]

Simultaneously, Henry's nephew, Henry Disston, son of William, searched for workers in Sheffield to supply the community with much-needed skilled labor. Of the 121 workers with property in Tacony in 1880, 54 were foreign-born. Foreign-born employees made up 70 percent of the skilled workforce. They included the recruited engineers William Boardman (age twenty-three) and Richard Seed (age fifty-eight). Almost all men over twenty-six (95 percent) were married, and their wives were classified in the census as housekeepers. Most of these men were skilled craftsmen. Twenty-eight of the workers were of apprentice age and were the children of foreign-born skilled workers. When Disston needed skilled labor, he usually found it more convenient to import workers from England than from the Laurel Street plant. So heavily did English culture permeate Tacony that teahouses were established in the community to continue to provide the custom of tea drinking.[27]

As the town of Tacony expanded, so did the factory. By the 1880s, machines had replaced many of the old hand processes. Handsaw teeth could be cut out at a rate of 500 a minute by a revolving cutter. Disston's Keystone Saw Works now took less than two minutes to complete a handsaw with 115 teeth. The circular-saw machinery was equally efficient. One man on a circular-saw grinding machine could produce six saws to every one produced by the old hand methods, and with fewer imperfections. Similar machines for hardening, tempering, and filing made Disston's plant one of the most productive in the world.[28]

President Rutherford B. Hayes visited Philadelphia in 1879 and chose to visit the Keystone Saw Works. He was escorted through the factory by Henry Disston's son, Hamilton. To impress President Hayes with the capabilities of the workers and the factory, Disston arranged to show him a flat piece of steel when he entered the factory. Two hours later Hayes received a new saw made from the steel he had seen at the beginning of his tour.[29]

This production pace and increased quality grew out of Disston's organization of the plant and the regimentation of the processes of steel- and saw-making. Henry Disston organized his factory into four basic divisions: steel, saw, filing, and jobbing. Tying together these divisions into a production process became one of Disston's early challenges in enlarging his factory. The saw works was central to all production schedules in the plant. Low inventories on specialty circular saws made the average wait for them about two weeks. Nevertheless, certain common handsaws were in stock and ready for sale. A saw order was received from the sales clerk, a man knowledgeable about what could be made at the plant, the time needed for its production, and the price of the item. The order was sent to the saw department, where the type and quantity of steel needed was ordered from the steel department. Because most files made at Disston were for use in sharpening saws as they were made, specialty files had to be in stock for finishing the saw order. All this meant that factory communications centered on the saw division, the needs of which determined the production of the other two divisions.[30]

Nevertheless, the steel works was the crucial beginning point for production of any saw. Without good steel, no quality saw would be made. Therefore, the final decisions about what steel was needed was left to the steel-makers, who operated independently of those who made the saws. (Similarly, file-makers made decisions on file-making independently of the saw works.)

Steelworkers had to have a knowledge of ores and metals, and because Disston found it difficult to find such workers in America he preferred to hire from the Sheffield area in England. Three techniques were used to make products from iron ore. First, workers would heat a mass of metal until it became workable, then the blacksmith would beat it into a shape using a hammer and anvil and later a steam hammer. The blacksmith building continued to operate throughout the Disston company's existence—proof of the importance and longevity of this technique.[31]

Second, the metal could be heated until it became fluid and then poured into molds or ingots. Because impure steel could not be used in saw-making, the crucible method for making steel was employed early by Disston. Involving small lead pots called "crucibles," this method was transported from England by skilled smithers. It produced purer steel and thus better-quality products. At Disston's company it gave way to the Bessemer process in the early twentieth century. Heat was also used during the hardening process. Oil was heated in large vats to the exact temperature needed to change the molecular content of the steel and increase the hardness of the final product. This procedure was important to saw-making, because the blades were cut from soft steel, requiring skill and concentration on the part of the worker.

Third, the machinist could cut, drill, chip, scrape, or grind the steel until it assumed the right shape and size. All these operations, except for shaping, were performed when the steel was soft. The machine process in making steel products was superior to the other processes in crucial respects. Introduced in the nineteenth century, it permitted micrometric accuracy. The drill, saw, and lathe were the machinist's first tools; later in the nineteenth century, the milling machine, planer, slotter shaper, and grinder allowed the production of more complex and irregular articles.[32]

Saw-making required skilled craftsmen who were knowledgeable about all phases of steelmaking and forging. The process changed little during the Disston firm's 115-year history. A simple handsaw required more than eighty operations. The steel was cut to proper form and shape using massive machine shears. The shears were hand-fed with such speed and precision that the process seemed automated. Each blade was stamped with a figure that indicated the number of teeth to the inch. Automatic toothing machines toothed the blade.

Soft and unfit for sawing, the blade was then sent to the hardening shop to be plunged edge first into a special hardening bath. This delicate part of the operation required special care until the steel had tempered. Excessive hardness was reduced by relaxing the molecular rigidity of the steel. The durability and elasticity of the blade depended on this operation.

"Smithing," the skillful hammering of the saw blade to straighten or flatten it, was the next operation. Smithers lined up beside the shop windows tapped steel to take out bumps and ridges. Blades were lifted to the light so that unwavering natural shadows pointed out where the hammer should fall. The regular rhythmic tapping of hammers sounded like the vibrations of some gigantic machine.

Across the room were grinding machines running the entire length of the building in an uninterrupted line. The handsaw was ground so that it tapered in thickness from tooth edge to the back and from the handle to the point. Natural sandstones, a major contributor to the deadly disease known as silicosis, were used in this process. Artificial grinding stones were not introduced into the plant until the 1930s.

Tensioning, the next step in the process of hammering, required considerable skill and experience, for this part of the process gives the blade the proper amount of spring or character. In a "fast" blade the metal needs expanding from the center. In a "loose" blade the metal must be stretched on the edge. A saw without the proper tension will not cut straight or may burn out from friction.

After the blade was ground a second time, "blocking" was begun by the best smither in the shop. Using lignum vitae blocks (hence the name "blocking"), smithers corrected the remaining slight irregularities. After blocking, the blade was polished and stiffened. Blades were then etched, and the setting of the teeth began. Skilled workers laid blades in special vises and set every other tooth with a tap of the hammer so the tooth extended out about half the thickness of the blade. This prevented binding. Sharpening with files followed, one tooth at a time. Finally, the straight-saw blades were sent to the handle department for assembly of the finished product and the circular saws had their blade balance checked. Using this manufacturing process, Disston & Sons produced every imaginable sort of blade for cutting wood or metal.

The jobbing shop produced tools of all types: bevels, squares, gauges, screwdrivers, levels, plumbs, miter rods, trowels, brick-pointing tools, hedge knives, wall scrapers, hedge trimmers, hand pruners, lopping shears, and porthole diggers. Wars brought Disston contracts for steel plate, rifle barrels, and other military necessities. From the Civil War to World War II, the factory was crucial to a winning war effort and was the world's leader in innovation and discovery in tool manufacture.

This description of the metal process illustrates the kinds of workers needed in Disston's file, saw, small-tool, and jobbing shops. For Henry Disston, the key to the process was having a large number of skilled workers.[33] The Disston machinists became skilled in determining the ideal "feeds and speeds" for each machine's operation. Their work was based on calculations involving a limited number of variables, all of them theoretically easy to discover, their interaction simple to work out by slide rule. Nevertheless, each worker kept his own personal "black book" with reminders as to the setting of his machine for specific orders.[34]

Smithers used up to twenty hammers to level and tension saws. Machinery could not replace human beings in this skilled market because no two saws contained the same steel with the same imperfections. Skilled labor was needed to finish saws, especially the large circular ones. Still another category of highly skilled labor consisted of a dozen or more workers who devoted all their time to machine repairs. Their responsibility was to make parts for the machinery used throughout the plant, and they could replace broken gears or rods within a few days.[35]

From the beginning the saw and file shops had been organized by function and separated into subshops. Under this system, workers were hired by and worked under foremen, each of whom was expert in a specific trade or skill. Augmenting the work of departments were foreman-led labor gangs that cleaned shops and moved raw materials, sandstones, and steel through the plant. It was typical of factories that such movement required a large supply of industrial labor. Payment, in cash, was delivered to the workers at the workplace in a wheelbarrow. As the factory became larger and production grew more specialized, subdivisions were created: the handle department, a circular-saw shop, a chisel-point shop (making removable and easily replaceable teeth for the end of circular saws, thus eliminating the need for resharpening), a bandsaw shop, a grinding shop for sharpening saws, a hardening shop, and a lumberyard.[36]

A series of buildings housed each of the saw-making operations; others were used for storage. In Tacony, the single factory became a manufacturing plant combining the rolling mills, the handsaw factory, the handle factory, the circular-saw factory, the butcher saw and trowel shop, the smithing shop, and the machine shop within the steel, saw, file, and jobbing divisions. Volume ordering yielded lower rates for raw materials and increased profits for the company.[37]

This vision of a manufacturing plant in Tacony would make Henry Disston & Sons the premiere saw manufacturer in the world. However, five years after beginning the project, Henry Disston became ill. He had suffered a stroke in late 1877, and after a partial recovery he went to Hot Springs, Arkansas, to recuperate. He returned to Philadelphia on March 9, 1878, and on the following evening suffered a second stroke, from which he never recovered. He died on March 16, 1878, in his fifty-ninth year. His eldest son, Hamilton, became president of the company.[38]

Obituaries described Henry Disston as a hardworking, persistent, and frugal man who merged his knowledge and skills as a machine-maker and saw-maker to produce the machines and materials needed to mass-produce saws. Taught by his father to trust in the family, he was willing to put responsibility for different factory operations in the hands of relatives. His ability to choose the right person for the job was one of his finest qualities. Thus, his death had only a minimal effect on the economic growth of the company. His most endearing quality, however, was the way he treated his workers, and his charge to his sons to continue to support the vision of a healthy and family-oriented community for his workforce.

CHAPTER

2

Paternalism and Nineteenth-Century Tacony, 1872–1900

HENRY DISSTON did not establish Keystone Saw Works simply to realize a profit. As his company increased in size, he used a considerable share of his profits to experiment with the idea of a planned factory town on the river. At the core of this endeavor was the factory owner, who was responsible for providing for the workers as well as for ensuring firm stability. Disston viewed the owner as bound together with worker and community in a mutual relationship—a paternalistic relationship—that advanced the company but also benefited everyone involved. This chapter focuses on Disston's paternalistic vision, beginning with a description of the community he created and speculating on the sources behind that vision.[1]

Henry Disston explained his belief in the mutuality of the pact between himself and his workers in a rare letter to his "fellow workers" on November 13, 1867:

> This [company and community of workers] is what I live for. We all ought to live to make each other happy. God knows the greatest desire of my life is to see all that I am connected with happy. And I believe to this day that there is not a happier or more contented family in the world. I say family because I consider you and myself of one and the same family. There has [*sic*] never been any wants that I could afford to alleviate but that I have endeavored to do so as I would my nearest kin. . . . The object of men and Boss should be mutual, the Boss to give all he can when times will permit, and the men under a close competition to be willing to

> help meet the market. . . . Whatever money I make is spent in improvements to facilitate us in putting goods into the market at such prices that we will have work as long as any house.[2]

When the company was small (1840–1855), Henry Disston was able to maintain close personal contact with all workers. A story told by Charles T. Gravatt in 1919 about how he got his job at Disston some forty years earlier illustrates how Disston's concern for workers became a part of his personal management style. Gravatt recalled that foreman Enoch Sinclair had not wanted to hire him. Founder Henry Disston disagreed. "Here is a man hard up. Give him a job." Sinclair objected that there was no work, but Disston prevailed. Gravatt described what happened next: "I was a knock-about at a small wage for a few weeks, then I was put in the circular saw department. I often think of the very pleasant relations that existed between the men and each of the members of the Disston family." Those good relations were cultivated by Disston's willingness to carry prospective good workers in hard times and by the workers' appreciation for Disston's trust.[3]

Disston worked alongside his employees and knew their names and family situations. His first employee was David Bickley, who remained on at the saw works for forty years, rising to a position as foreman of the long-saw department. In 1890, when he retired, Bickley still spoke admiringly of how Henry Disston toiled side by side with him in the heat and cold of the 1840s to make saws. The idea of worker dignity and sharing tasks was left over from Disston's childhood in England. He had been born into a family in which each member had participated in the family business. His grandfather owned Nafford Mills, and his father had worked there at an early age. Both believed that humane treatment of workers did not necessarily mean low profits. Henry Disston worked for his father and undoubtedly inherited from him this respect for workers.[4]

However, as Disston's firm prospered and grew in size, these informal arrangements proved inadequate. This became obvious when the firm began its move to Tacony in 1872. The move was prompted by the need for expandable land space and a stable, skilled workforce that would permit economic growth of the company. Two sites and the size and complexity of the operation demanded new organizational arrangements. A system of work had to be found that responded to the problems then developing in large-scale factory operations: absenteeism, tardiness, labor turnover, and labor disruptions. Rather than adopting authoritarian or bureaucratic methods of control, Henry Disston opted for a more paternalistic approach designed to bring together family, factory, town, and workers.[5]

First, Disston adopted a gamut of special provisions that benefited his workers and far transcended a simple exchange of labor for pay. Landownership was a key factor through which Disston planned to make the Tacony community and his factory one entity. Around the factory, he purchased 300 acres of land which he leveled in order to build streets and more than 500 homes. He also provided such common facilities as schools and a firehouse. George Smedley Webster planned the community by surveying the land and dividing it into salable lots. At Disston's request, these lots were spacious enough to permit the building of twin homes while still giving the occupants ample light and air.

Next, Disston began a building and loan association in 1874 to enable workers to purchase their own homes. He appointed himself treasurer and owned most of the stock. The association, which provided downpayments and second mortgages to the people who worked at the Disston firm, grew and prospered. When its capital assets reached the legal limit of the law in 1892, the Disstons were forced to establish a second building and loan association. A congressional report on Disston's associations referred to them as "models of their kind; among the strongest and most successful in the country."[6]

The Disston Real Estate Sales Book No. 1, kept at the factory, has a record of all properties sold to workers by the Disston family in the years between 1872 and 1896. A worker had a number of options if he wanted to move to Tacony to work for Disston. He could purchase a lot for $200 to $600, depending on the distance from the factory, and build his own home; he could rent three to four different-size houses for $15 to $25 a month; or he could buy a home at prices from $2,100 to $3,000. Rent or loan payments to the building and loan association were collected on the basis of the worker's ability to pay. Community folklore states that there was never a foreclosure on a Disston home. Disston's terms for loans and rents were considered reasonable and fair by the standards of the day.[7]

In addition to housing, Disston made other provisions for the community designed to make life more livable for workers and their families. The community park is one example. Why a park in the center of Tacony? In an area bordering on woods and farms, picnics and outings in the countryside were only a few steps away and common events for those who lived in nearby Holmesburg. But Tacony needed a park to symbolize the orderliness and culture of a respectable, civilized community distinct from the surrounding land patterns of farms and semi-wilderness. It also became one of Tacony's most prized attractions.[8]

The Music Hall was another community facility sponsored by the company. Like the park, it also projected the image of a civilized community. Built in 1885 west of the Trenton-to-Philadelphia railroad tracks in an area of open farmland, the magnificent three-story building with ornamental front facade and fancy brickwork stood out like a beacon, inviting both strangers and residents to a center for social and cultural community events. It contained three storefronts, which eventually formed the nucleus of the town's new commercial center. The second floor featured a large dance floor and space for an orchestra, and the third floor accommodated three meeting rooms, two of which were used by the benevolent societies as meeting places, while the back room housed the community library and scientific society. The large colored-glass windows, embossed tin ceilings, and library room remain today in the refurbished 1885 building.

Crucial to the financial success of the Music Hall was the support of the Disston family. Jacob Disston arranged financing and volunteered rent payments for the scientific society and library. The Disstons believed that the availability of books and journals, such as *Scientific American,* was important for the continued personal growth of their workers. The Music Hall, the park, the supply of pure water, and the spacious lots for housing were essential for attracting skilled workers to Disston's family-oriented, civilized country village.[9]

For Disston, religious freedom was also essential to developing a peaceful industrial community. Religious bias by factory owners in small towns could limit the available workforce and create disharmony between ethnic and religious forces in the community. Because Disston's skilled workers belonged to different religious groups and sects, Disston believed the community should support religious diversity. To achieve this end, the Disston family sold land to each of the community's churches. Henry Disston was himself a deeply religious man, a practicing Presbyterian who was broadminded enough to respect the religious beliefs of others. The attitude of the family in projecting religious tolerance did much to promote tolerance in the town itself.[10]

Another provision for the workers involved leisure-time activities. It was not unusual for the Disstons to grant their employees a day off with pay and present them and their families with an excursion down the Delaware River. A planned yearly excursion by the company's Keystone Beneficial Association to Riverside Amusement Park raised $535 to augment the dues. The Beneficial Association's funds were deposited with the Keystone Saw Works, whose officers supervised and determined the appropriateness of the payments. The fairness of the company was rarely questioned by the employees. Special citywide events also prompted company generosity. In 1876 Henry Disston presented each employee with tickets to the Centennial Exposition at a cost to himself of $500. And, as was the custom of many manufacturers, Christmas turkeys were presented to workers with families.[11]

A "benefits package" was arranged to protect the workers from wage loss because of illness. Begun by Henry Disston himself, the Keystone Beneficial Association provided illness benefits of $5 a week for its members, who paid $1 a month for membership. In addition, a death benefit was provided for workers' widows. By 1918 the association had a membership of 594 workers and met monthly in the Tacony Trust building. The 50-cent fee paid at each meeting ensured them of a $100 death benefit and $50 monthly payment in case of long-term illness, their only security against catastrophic illness.[12]

When such a catastrophe did occur, Disston was equally active in setting aside sums of money yearly for other than real-estate activities. He built a private dispensary at the Laurel Street plant, where all the poor of Kensington, as well as Disston's workers, could get free medical treatment. During severe business depressions, when unemployment was widespread in the neighborhood, he maintained a soup kitchen for the hungry and the jobless.[13]

Disston believed that protecting his workers from layoffs was also his responsibility, but he did not ignore economic reality. During a mild economic slump in 1867, he spoke of the special relationship between sales and wages for workers. He could protect his workers from unemployment only if they trusted him to pay well when business was good and to reduce salaries during poor economic years:

> Then again I say, let us put our best exertions together and see if we can keep full time. . . . I only ask you to help me make some small portion of the reduction and the reason I shall ask it is because I know it will be to your benefit in the long run. We can keep up the prices but be short of work, and short time will pay neither you nor me. Now please allow me to thank you for all your past and present evidence of good wishes, and I assure you that my greatest exertions shall be spent in trying to make both you and your dear family as happy as possible, and at any time when you are in trouble, sickness or distress, I shall certainly take great pleasure in trying all I know to make you happy.[14]

These provisions and efforts to protect the workers put Henry Disston in a position to control the behavior of his workers not only in the factory but also in the town. Clearly some of the provisions aimed at protecting the community required controlling the life-style of the community. Deed restrictions that provided for a family-oriented living space in the community also controlled workers' drinking habits, controlled industrial growth in the town, and limited political activity in the town by not permitting a town hall. (The first large meeting hall in the town was the company's cafeteria, built in 1919.) Control of politics meant providing meeting places in the factory.

The economic institutions and power structure were all influenced by the family. Henry's eldest son, Hamilton, second son, Jacob, and Magistrate Thomas South were the power brokers of the community in its early decades. Magistrate South, with Hamilton Disston's support, controlled the justice system while doubling as the real-estate agent to whom the company referred those seeking to rent or buy a home in Tacony. Financing for the homes was arranged by the current secretary of the Keystone Saw Works, Jacob Disston, who doubled as president of the Tacony Trust Company. A newspaper, appropriately called the *Tacony New Era,* was sponsored and controlled by these financial institutions and the Disston company. The few who might disagree with Disston policies or oppose the values expressed in the deed restrictions had no means of addressing their grievances except to leave the community. Disston's emphasis on human values and close association of family members with the company developed a trust between worker and owner that lasted for more than one hundred years.[15]

Second, as part of the paternalistic pact with his workers and the community, Henry Disston considered it his duty to insulate both from the world. An informal decision not to permit bells in the church steeples left the town and factory free from the rowdy behavior of the volunteer fire companies that had disrupted his Kensington factory site.[16] The deed restrictions placed on these 300 acres of designated residential building lots were clear and were enforced:

> No tavern or building for the sale or manufacture of Beer or Liquors of any kind or description and no court house, carpentry, blacksmith, currier or machine shop, livery stables, slaughter houses, soap or glue boiling establishment or factory of any kind whatsoever where steam-power shall be used or occupied on the said lots, tracts or piece of land or any part thereof.[17]

These restrictions were meant to make the neighborhood more livable for everyone and to provide for what Disston viewed as a better life. Smell and noise abatement and a ban on saloons were features of few Philadelphia communities, but Disston and his company stood to benefit from these rules as much as the workers did. "No taverns" meant that workers were sober for work; no bells in churches limited factory disruptions; the absence of steam engines meant less competition from other factories for the community's meager workforce; and the ban on stables, besides eliminating smells, limited the distance workers could live away from the factory, forcing newcomers to the Disston Works to buy land close to the factory—land owned by the Disston family. Moreover, elimination of these urban nuisances enhanced the value of the land held by Disston.

Perhaps the most far-reaching element of Disston's paternalistic program was apprenticeship (see Table 2). It not only provided skill training for new workers but also protected jobs for the sons of Disston's workers, who were by far the largest group of apprentices at the factory, and it allowed Disston to control the behavior of the young men in the community. Because Henry Disston believed so much in the value of apprenticeship, he required each of his sons to apprentice in one of the shops.

Apprentices generally began working between the ages of 9 and 17. The Disston contract contained provisions that were standard for most apprenticeship contracts:

> The apprentice doth covenant and promise, that he will serve his master faithfully, keep his secrets, and obey his lawful commands . . . that he will not contract matrimony within said term—that he will not play cards, dice, or any other unlawful game, thereby his

Table 2. Apprentices and Adult Disston Workers, Tacony Factory, 1880 Census

	Ages of Apprentices* (N = 28)				Ages of Adult Workers (N = 93)		
	Under-10	11–12	13–15	16–17	18–25	26–39	40+
File Shop	2	3	8	7	21	9	11
Saw Shop	1	2	1	1	8	11	11
Machinist	0	0	0	2	0	4	6
Laborer	0	0	0	0	0	3	5
Engineer	0	0	0	0	1	0	1
Steel melter	0	0	0	0	0	0	1
Office boy	0	0	1	0	0	0	0
Night watchman	0	0	0	0	0	1	1
Totals	3	5	10	10	30	28	35

SOURCE: U.S. Census, 1880.

*Most apprentices lived in the same house as an adult worker.

> master may be injured—nor haunt ale-houses, taverns nor play horses, but in all things behave himself as a faithful apprentice.[18]

In return, the contract stipulated a four-year period of apprenticeship at one-fifth of what adult workers earned, but it usually guaranteed a steady wage and possible advancement in the firm. Disston's benevolent apprenticeship practices gained for him a reputation among his workers that would become, over the next thirty years, part of the folklore of the town. It was comforting for a town's people to feel that someone cared about them and was providing training for their sons. Such feelings helped to promote stability and discourage labor unrest, but they also served to narrow attitudes by focusing attention on local community life and instilling suspicion of all outside influences. Disston family descendants would become the main beneficiaries of these sentiments.

As for the Disston sons, the practice of apprenticeship continued into the 1940s, when William Leeds Disston became the last of the family to learn the saw trade from the men. However, this long practice of apprenticeship ensured that each future boss from the family had a first-hand knowledge of the plant's operation when it came time for him to assume a position of authority within the firm. It therefore would not have been unusual to see Henry Disston or another family member in the shops on a daily basis addressing workers by name. There developed between the owners and the workers a mutual respect for their respective roles in the company. Inside the factory, Disston encouraged each foreman to be a paternalistic father to his workers. This attitude prevailed into the 1950s, as demonstrated by smither William Rowen, recalling how his own foreman, Johnny Southwell, was like a father to him. Rowen could still recall the Disston family members coming through the shop and addressing him by name. Disston envisioned his company as an organic society in which all workers had a specific place and function, all bound to one another in a network of reciprocal obligations. The workers were guaranteed a steady employment, and the Disston family gained a community that was loyal to the firm. This unwritten agreement was meant to create a workplace free from labor unrest and a neighborhood that reflected stable family values.[19]

There are no writings from Henry Disston himself and few descriptions by others that indicate the motives that sparked the Tacony experiment. Other than a concern for more space to expand the factory and an expressed desire to relieve the crowded living conditions of his workers, he did not indicate what he intended to achieve in his model town of Tacony. We will never know for sure where Disston got his inspiration for the system he adopted in his town on the river. The best we can do is to speculate from the information we do have: the things he valued, his life-style, family influences, and those influences of the world in his day that might have been instrumental in shaping his ideas on paternalism.[20]

Arguments over the rise of factory capitalism were frequent in the pre-Victorian England of Disston's childhood. As an apprentice mechanic with his father in the area around Sheffield, an industrial center, young Henry undoubtedly heard the debate over the effects of commercialism and the breakdown of paternal relationships then pervading all segments of English society. Disston was exposed to what one historian has described as an English society that saw every facet of life permeated by paternalistic thought. Thus, Disston grew up in an atmosphere where "no social outlook had deeper roots and a wider appeal than did that which twentieth-century historians call paternalism."[21]

Religion was yet another potential source of the Disston family's awareness of paternalism. Henry Disston and his wife, Mary, both had strong religious beliefs, and as Presbyterians they were active on behalf of Philadelphia's Social Gospel movement. Toiling in the working-class neighborhoods of the city, Presbyterians were attracted to social work with such vigor that the Philadelphia Presbytery grew 38 percent in the latter half of the nineteenth century. By then, for Presbyterian elite like the Disston family, benevolence was a way of life. Evidence of this was Henry Disston's and shipbuilder Charles Cramp's sponsorship in the 1870s of Beacon Presbyterian Church, which ministered to more than 1,500 people through an aid society, a dispensary, and a working people's institution.[22]

For most Presbyterians, however, charity was one thing but social welfare was quite another. Throughout the years, they scorned social legislation as irresponsible and instead advocated a philosophy of self-help based on individual responsibility and a virtuous life. The duty of well-off Christians was not simply to offer handouts but rather to provide the opportunity for people to reform themselves.

Mary Disston, a devoted Presbyterian, was intensely interested in promoting religious institutions and a humane social order. Born in Atlantic County, New Jersey, she was the daughter of Jonas and Ann Steelman. Her ancestors could be traced back to before the Revolution, and her great-grandfather recast the Liberty Bell as a workman for Pass & Stow Company. Mary supported Henry's out-of-factory benevolence and still had sufficient time to be mother to eight children. She was involved in all sorts of public charities and provided funds for soup kitchens and children's shelters. The death of a daughter at an early age provided the impetus for Mary to donate funds to build the Tacony Presbyterian

Church. To this day a plaque in the rear of the church reminds the congregation of her generosity and feeling for her child.[23]

Mary's ability to complete difficult projects without Henry's help is best exemplified by her interest in Atlantic City. After a visit to the shore, she commissioned a friend to build a summer home on Atlantic Avenue above Illinois in 1872. Mary planned and designed what she felt would be a family summer home, but neither she nor Henry saw it until it was completed and fully furnished. Mary then brought her husband to the house for dinner as a surprise. Henry liked his new vacation location, and in 1873 opened the first steam sawmill in Atlantic City. By 1879, through the encouragement of Mary Disston, Henry Disston & Son had won the contract to build a sixteen-foot boardwalk the length of the city.

These limited accounts of Mary's role in family matters suggest that she was a capable person in her own right and unafraid to manage things on her own. Her personal strength accounts for the continuation of Henry's paternal practices in Tacony after his death in 1878. Between 1875 and 1895, through her control of the vast Disston property holdings in Tacony, she was responsible for giving land for schools, churches, and a playground. Subsequent to her death in 1895, the people of Tacony named the newly constructed public school in her honor in 1900.[24]

The Disston system seemed to work in this small-town setting. There is little evidence that workers resented the Disstons' influence on their lives. Labor relationships were relatively peaceful throughout this period—a time of strikes and violence throughout America. There were only two instances of threatened strikes, and one short strike, but no violence against the factory or family.[25]

In 1877, as labor organizations burgeoned in Philadelphia, one of Disston's men tried to organize the workers at the Laurel Street plant. The Disston management sent word to the workers that it was "considered injurious to the interest of the company to retain" the man. He was quickly discharged for disloyalty. In sympathy, some of the men followed the organizer off the job. For the next two days, little or no work was done at the plant as labor organizations in the city protested. Disston hired new workers to take the place of those who had left. Within a few days the regular workers had returned, and all except those who were most vocal against the company were rehired. This first brush with unionism left Disston relatively unscathed. The later move to Tacony further muted labor dissatisfaction.[26]

The bond between company and community in Tacony survived waves of social and political change. Nationally, labor discontent swelled the ranks of the Knights of Labor from 50,000 members in 1884 to more than 700,000 in 1886. An alarmed Rutherford B. Hayes prophesied that "free government cannot long endure if property is largely in a few hands and large masses of people are unable to earn homes, education, and a support in old age." In Philadelphia, agitation by the Knights of Labor that initiated a decade of unrest for local manufacturers began in 1882.[27]

The first labor disturbance at the Disstons' Tacony plant occurred in early May 1884. The disagreement began when the company released Joseph Broomhead, a popular supervisor and a seventeen-year employee, because the Disston management was disappointed in his work performance. A local newspaper story, reported Broomhead's contention that the firing was "without apparent cause." Horace C. Disston (1855–1900), Henry's third son, who was running the company in Hamilton's absence, hired S. T. Williams of the Albany Iron Works to take his place. In early April, Williams sent his nephew, George Thompson, to take charge until he finished his work in Albany. Thompson's first act was to fire two men and employ boys at lower wages. About the same time, letters were received in Tacony from Trenton, New Jersey, and Johnstown, Pennsylvania, along with an issue of the *Troy Observer,* all containing criticisms of the way Williams managed workers. He was charged with "brutality to the men under him and introducing Swedes, Hungarians and in a few cases Italians in the mills under his charge for the purpose of reducing wages."[28]

The workers from the rolling mill demanded a meeting with Horace C. Disston and confronted him with the information on Williams and an ultimatum: "Unless Williams was discharged they would all leave in two weeks." Disston subsequently closed down the rolling mill, throwing fifty-one men out of work.

James F. Ryan, chairman of the workers' committee, went to the city's saw manufacturers during the next week to inquire about jobs for the men. No company in Philadelphia could accommodate such large numbers on such short notice. The Knights of Labor promised assistance, but Tacony was far removed from their power bases in the more populated industrial sections of the city. Meanwhile, the company took advantage of the mill closing to make improvements to the building. It was having it both ways—initiating improvements and punishing militant workers. After two weeks, the mill was opened again and Horace Disston was ready to reinstate the fired workers, "with the exception of certain leaders." Believing that a few leaders had misinformed the men that wages would be lowered if Williams's improvements were implemented, Horace assured them that the gossip about wage cuts was untrue. The men returned to work under Williams at a 5 percent increase in pay. The subsequent work of Williams over the next ten years vindicated the Disston position. Williams became a valued member of the management team and was eventually accepted by the workers, who found him to be a competent, fair, and respected boss. Williams's retirement in 1894 from Disston Saw Works was marked with warm regards from all.

In 1886 a five-week strike, the longest of the century for Disston, occurred over a wage cut. In January 1885 the melters in the Henry Disston & Son steel mill accepted a 10 percent pay cut because they were told the company lacked orders. They were promised that the cut wages would be restored when business picked up. On June 23, 1886, the melters walked out on strike, claiming that business had improved and demanding restoration of their wages. The company responded with a 5 percent offer, claiming that Disston was already paying higher wages than any steel producer in the state. To prove this, the company offered to pay for a committee (two representatives from the steel melters and one from the company) to visit steel mills in Pittsburgh to compare wages. The steel melters refused, stating that steel was made differently at Tacony and that "no comparison could fairly be made." The company considered this an act of bad faith and appealed to the striking workers from departments other than the steel mill to be the judges. These men were more willing to listen, since they were losing wages and had nothing to gain from the settlement. They formed a committee of ten, which went to Pittsburgh and verified the company's statements about steel wages. The strike was broken, and, according to one Disston superintendent, the men returned to work, concluding that there "was no just cause for the recent strike in that department."[29]

These three fragmentary accounts of labor dissatisfaction at Disston & Company demonstrate the relative peace that existed in Tacony during what can be described as otherwise turbulent and violent years. Contrast these nonviolent activities with events at the Amalgamated Association in Pittsburgh, where on July 6, 1892, the Homestead Mill clash between 300 Pinkertons and the workers resulted in six deaths and scores of injuries. Violence in the years that followed led to a concerted effort on the part of Pittsburgh steel plant owners to establish nonunion workforces. By 1900 the antiunion stance of the powerful Carnegie Steel Company had driven unionism from the steel mills of Pittsburgh.[30] In Tacony, the Disston family won the day not by violence but by job practices that were fair according to the standards of the day and a paternalism that promoted trust between factory owner and the community at large.

Traditionally, saw-making skills were passed from father to son, and it was common for Disston workers to bring their sons, grandsons, nephews, and sons-in-law to work at the factory. Once a family settled near the plant, all its men and older boys sought jobs there. As the years passed and Tacony became famous for its craftsmen, employment at Disston became almost a birthright. The paternalistic company could gain greater control over the workers' lives by allowing family members to fill factory vacancies, by arranging the purchase of land for churches, and by establishing savings institutions so workers could purchase homes. This promoted a common vision, between worker and owner, of mutual exchange: The worker contributed his labor, and the owner benevolently protected workers' interests; the owner could expect few labor disturbances, and the worker could expect a steady income and a better life. The mutual acceptance of these roles was what made paternalism so successful in Tacony's early years.[31]

Typical of Henry Disston's contented and stable skilled workforce was the Arnold family, pictured in the company's monthly paper, the *Disston "Bits,"* outside their Tacony home in 1921. The "*Bits*" took the opportunity to point out that the three brothers—John, George, and James Arnold—and their sons, had been Disston saw-makers for an aggregate of 256 years. Such family ties go back to the firm's early days.[32]

The nature of the workforce also discouraged unionization at Disston. It was difficult for these skilled workers to consider membership in the Knights of Labor because the labor movement was committed to equality in the workforce, elimination of craft distinctions, and consolidation of workers in one national labor movement. Disston's skilled craftsmen opposed both possibilities. They wanted to maintain their status in the factory and were opposed to a national union that could dictate to their local community.[33]

Clearly, paternalism was the common way of managing business in Henry Disston's day. Considered by some as a cost-saving, rational business practice, and by others as an extension of patriarchal family authority from the home into the factory, it represented a "reconstruction of the father's domain over a set of industrial children." Evidence of factory paternalism abounds, but the paternalistic arrangements in places like Lowell or Manchester, in the rural North and the post-Reconstruction Carolinas, were unlike those in Philadelphia. Within Philadelphia itself, the Disston model differed from the system established at Stetson Hat Company where landownership was not a major factor. The paternalistic experiment that was most similar to that of Henry Disston was the one imposed by railroad magnate George Pullman in Chicago. Both were based in virgin communities developed by the company and located within the boundaries of a large urban city. Both men based their paternalistic plan on landownership and control of the town's political and social activities. Both companies were the largest national firms in their particular product line. Both companies required skilled workers to make their product. And finally, both companies had a dynamic leader that dominated the management of the company. Therefore, a comparison of the Tacony experiment with George Pullman's utopian factory community in Chicago may shed some light on the success of Henry Disston's paternalism.[34]

In the 1870s Pullman decided to move his factory to the open lands on the south side of Chicago. He bought large quantities of land and hired an architect to design his town and houses so that they had "artistic character." Stores placed in an arcade building, a park, an athletic club, and an artificial lake arranged to set the residential section of the town apart from the factory were all built by the company. None of the houses was sold by Pullman—they were all rented. Land was purchased between Pullman's new community and Chicago proper to ensure that the town would remain isolated from the problems of the city. The goal behind the "Pullman System" was harmonization of a town and factory with the natural open space of the environment.[35]

Pullman, however, referred to his paternalistic town as "strictly a business proposition." His policy of increasing rents further frustrated workers, who resented his making a profit on every aspect of their lives. Nevertheless, Pullman expected his town to attract and retain a superior type of worker, who in turn would be "elevated and refined" by the physical setting. Contented employees would naturally lead to a "reduction of absenteeism, drinking, and shirking on the job." Ultimately his workers would be less susceptible to the exhortations of "agitators" than the demoralized labor of the slums. The result for Pullman would be increased output.[36]

Disston's similar ideas for his Tacony attracted much less national attention, but he was able to avoid labor unrest and promote worker loyalty. George Pullman, on the other hand, was dismayed when his town became the center of labor and social disturbances during the strike of 1894, and his experiment was deemed a failure by the press, who referred to it as "utopian paternalism gone sour" and an "unsatisfactory . . . domination of community interest by . . . industrial authority."[37]

Far to the east, another scenario was being played out in Pennsylvania. In 1886, while the Haymarket bombing in Chicago was generating a crisis atmosphere in the nation, and as Philadelphia marked the end of four years of conflict between the Knights of Labor and Philadelphia manufacturers, Pennsylvania's secretary of internal affairs surveyed the state for a successful, ongoing, and benevolent relationship between worker and manufacturer. He found such a model in Tacony, Disston's newly settled industrial community. In 1887 the secretary's report on industrial statistics included a detailed description of Disston's firm, whose practice of dealing fairly with labor had minimized conflict and created a model manufacturing village. Disston was the hero of the report, a model for other manufacturers. Tacony was held up as the success story of a management system that recognized the need for workers to have a fair wage and steady work, opportunities for mental and moral training, and healthful physical surroundings for their families:

> A visit to this well ordered healthful village, a peep into the homes of the working men, an inspection of the factories, the evident attention to light, air and other sanitary arrangements, the fair treatment which the employed has always received, these things must convince the visitor that at Tacony sure progress has been made in solving "the labor question."[38]

Besides documenting the beneficial working and living conditions of the employees, the report emphasized Disston's humane treatment of workers. The advantages to Disston included not only labor peace where wage issues were concerned, but also an easing of tensions when labor-saving machinery was introduced into his factory. Although Disston's workers had to be convinced of the value of a novel technology, "none of [them] ever destroyed the newly-introduced machine."[39]

The beauty of the Disston Tacony model, the secretary pointed out, was that it fostered a natural cycle of growth and profit. Mechanization increased production, which fit nicely into post–Civil War economic expansion and resulted in a demand for more saws. This in turn perpetuated the cycle of guaranteed work with few wage cuts for the workers. Steady wages produced a stable, family-oriented, optimistic working-class community where men worked and women stayed home with the children. Production increases also benefited society by decreasing prices. The secretary ended by exhorting his readers: "Let the despairing go there if they wish to revive their hopes concerning the future of the working class. As demonstration is better than theory, study the history of Mr. Disston's enterprise and the vision of happier times will appear to you."[40] Begun at a time when the ideas of paternalism were popular in America, Disston's enterprise still embodied these older values.

The report of the secretary of internal affairs encouraged Disston's widow Mary to continue to build homes and provide services for their industrial-utopian community. With Mary Disston in control of the Tacony land, and her sons running the factory, the family prospered. Upon Mary's death her estate remained intact, benefiting the workers and paying her grandchildren dividends as long as they lived. The death of the last Disston named in her estate, in 1943, resulted in the sale of the

Disston houses. As provided in the will, the houses were first offered to the workers who lived in them at reasonable prices. Mary also stipulated that each house sold be first put in good repair, with a new roof, a bathroom, wallpaper, and a kitchen if necessary. Every house was purchased by the worker living in it, with one exception, which was immediately purchased by Catherine Seed, rent collector for the Disston Estate. This generous treatment of the workers by the family at this late date cemented the Disston reputation for benevolence in the minds of Tacony residents.[41]

The growth of the community in the late nineteenth century was phenomenal, mirroring the growth of the saw works. The city property atlas of 1862 indicates only 22 homes, scattered along Tacony's riverfront. In 1876, just four years after Disston raised his first factory building, the riverfront was crowded with homes, and there were 42 twin homes and 12 row homes, newly constructed, to the west of the Philadelphia–Trenton railroad tracks. By 1894 the village had become a small town with well over 500 homes. These were, with a few exceptions, twin houses. Disston's belief that children and families needed light and yard space had been put into practice. Using a sliding scale, Mary Disston rented a two-bedroom house for $8 a month and a five-bedroom house for $15. The schoolhouse, churches of different denominations, a bank, a music hall, a firehouse, a police station, waterworks, and a park were there for the employees' use and protection.[42]

The success of Disston's utopian industrial town and the failure of Pullman's development of a similar town within the limits of a large city raises questions as to how such similar schemes could have such different results. There were three elements in Disston's plan that were not in Pullman's plan. First, Disston's practice of having all the male children apprentice in the shops allowed a direct line of communication between the workers and the family. Once the grandchildren began apprenticing in the company, there were as many as six Disstons in the various shops of the company. Many of these same family members, especially the grandchildren, settled in Tacony between 1900 and 1915. It was common practice at Disston for skilled workers to confer with upper management and family members. Clearly, both formal and informal lines of communication connected the workers to the Disston family, and the town to the factory.[43]

George Pullman, on the other hand, seldom visited his town after 1890. In the year before the 1894 strike, Pullman visited his workers' town but six times, and those were mostly ceremonial appearances. Historian Stanley Buder points out that the Pullman Company suffered from a lack of formal lines of communication between employer and employees, which meant "that the workers did not approach management until cumulative grievances etched deep suspicions which predisposed the men to a hasty and emotional reaction."[44]

Second, Disston and Pullman differed in their view of how residential housing should be administered. Pullman believed that housing, like the factory, was a money-making proposition, so he only rented his properties and continually raised rents, which caused great unrest among his workers. Disston viewed housing as a means of attracting a workforce that would be loyal to the company. He seldom raised the rents, and he allowed mortgages to go unpaid during periods of economic distress.[45]

Finally, Henry Disston's concern for his workers' personal lives, as exhibited by his reluctance to order layoffs and by his liberal housing provisions for his workers, was legendary in Tacony. Disston's concern for people can be contrasted to Pullman's belief that aesthetics had a virtue and function of its own, which he called the "commercial value of beauty." Pullman came to rely on the beauty of his town as the means of influencing workers, whereas Disston relied on personal contact. The image among the town's people that George Pullman was "hostile to labor" further complicated his ability to influence his workers. On the other hand, Tacony residents remained loyal to the Disston family and firm into the 1950s. People in Tacony believed that the Disston family was interested in them, while the people in Pullman saw themselves as mere puppets in a scheme to make George Pullman and his stockholders rich. These differences in perception allowed Disston to settle work disruptions and keep labor unions from organizing, while Pullman experienced labor difficulties that eventually led to the demise of his experimental industrial town.[46]

Although Henry Disston never wrote a book on business principles, a chronicle of his daily pursuits illustrates a four-cycle formula for success. Simply stated, he decreased cost and overhead, enlarged the market for his product, avoided labor disputes, and accumulated capital through a process of reinvesting money in the company and land. Successful experimentation with steel formulas, file-making, and the bandsaws used in making saw handles increased the number and quality of the saws produced. The differentiation of labor and the establishment of departments to make specific saw parts—a rolling mill to make flat steel and other shops for tooth-cutting, handle-making, and finishing—decreased unit costs.

Yet the Disston family's accumulation of wealth was accomplished without embittering the workers in the community. Treated well, those living in Tacony owed a debt to the Disston family for their support of almost every new institution in the community. Land was sold at reasonable prices to the workers yet brought great profits to Disston. As the community's population increased from a few hundred in 1871 to approximately 12,000 in 1910, each new house made the remaining land more valuable. This was consistent with the claim of the progressives of that day that in the new industrial society there would be no losers. Disston's purchase of land for the good of the employees was one of the shrewdest real-estate deals of his day. Disston's formula for success underlined what the interior secretary's 1887 report wanted Pennsylvania industrialists to memorize—namely, that looking after your workers in a benevolent way resulted in handsome profits. It was good business to be a "good" boss.[47]

The success of any industrial paternalistic community can also be measured by the local leaders' feelings about themselves and the controls exercised by the factory over their lives. In 1906 the leaders of Tacony published a booklet about their town. Released on May 30, the "Souvenir Program of the Celebration of Tacony" had pictures of local institutions and emphasized recreation, the local newspaper, the churches of every Christian denomination (there was no Jewish congregation), the new free library, the new post office, and the two local hotels. In a statement laden with middle-class boosterism, the shopkeepers and small businesses declared:

> Tacony is a paradise for the working man of moderate means. Here he can enjoy the refreshing comforts of good air, pure water and healthy surroundings. While Tacony is not what might be termed an aristocratic suburb, it nevertheless has a great many handsome residences, and the pairs of pretty homes which are continually being erected add still more to the place, and give it an atmosphere of comfort.[48]

This idyllic description of Tacony was remarkably similar to the one issued by the Pennsylvania secretary of internal affairs some eighteen years earlier. It would seem that the Disstons' paternalism was valued by those who experienced its benefits. Henry Disston's vision became their way of life, and long after his death the people of Tacony agreed with him that social control of the whole community is acceptable if the leader is benevolent and there is a bond of mutual respect.

CHAPTER

3

Henry Disston's Management Style and the Accumulation of Wealth, 1860–1878

HOW DID HENRY DISSTON & SONS acquire the enormous capital needed to build an industrial community? In New England earlier in the century, investors had banded together to accumulate the capital to establish Lowell, Massachusetts, and Manchester, New Hampshire. They had purchased land and developed their communities away from the city. Henry Disston used no such scheme, preferring to finance the venture himself. When he died in 1878, the family had little idea how much wealth he had accumulated or even where the wealth was invested. It took lawyers and accountants seven years to ferret out exactly what Disston owned and to total his assets for probate. The final figures were impressive.

Disston's wealth was distributed among land, mortgages, bonds, stocks, and the factory. The factory itself—buildings and machinery—was valued at $931,435.50. Disston's vast holdings in Atlantic City were valued at $40,477.50, which included three wood mills, land, and vacation cottages. His real estate in Tacony was even more extensive. Total expenses for developing Tacony and the factory were $106,593.98 in 1872 and $106,464.26 in 1873. This included plant buildings and equipment and the purchase of farmland from Christopher Eastburn, George Hammerly, and James Robinson. Between 1873 and 1895, land and homes were sold for a total of $701,368.87, and rents collected amounted to more than $300,000. The remaining land was left to Mary

Disston, to be sold over the next fifty years at many times that figure. In addition, Mary had extensive landholdings in Tacony in her own name.[1]

The company books show that Disston intermingled his company resources with his real-estate investments. Repair crews from the plant maintained his Tacony homes. Records on his Atlantic City properties were kept by his factory accountants, as were real-estate records for Tacony. It appears that Disston used capital accumulated during the Civil War to purchase real estate, although the mixing of investment and factory accounts makes an exact analysis difficult.

It is obvious from Disston's will that land was his major investment. His penchant for landownership dated back to the time he was evicted from his first saw shop, but land purchases often left him cash poor. This was the case in 1872, when he wrote a friend: "I am sorry to be compelled to inform you that I have no money to spare or put out at interest or otherwise I have all my spare money forecasted for the next 2 years to come."[2]

Yet much of the money made at Keystone was reinvested in the factory. Disbursements in 1867 for a new factory building ($11,539.78) and a large steam hammer ($7,500.00) account for no dividends being paid that year. A review of Private Ledger No. 1 of the Disston company for the years 1867 to 1869 identifies no monies being paid in dividends, only salaries for the members of the family who worked at the firm. The company continued to reinvest in itself until 1877, when dividends were declared for the first time.[3] Even then, dividends were small, though business was brisk, as a list of dividends in this period suggests:

1877	$20,000.00	1881	$24,162.50
1878	$4,997.84	1882	$11,059.00
1879	$4,054.50	1883	$14,453.91
1880	$20,635.00	1884	$4,304.30

In 1885 the dividend was $3,051.22, divided into five parts: Mary Disston (wife) got $1,017.07, while Henry's half-brother Samuel and sons Hamilton, Jacob S., and William, and the estate of Albert H., got $406.83 each. These dividends were small compared with the expenses inherent in developing a community like Tacony. Therefore, one must assume that most of Disston's real-estate investments paid for themselves and eventually became very profitable.[4]

In addition to real estate and the wealth generated by the business, Disston had control of the community's utilities. By 1878 a water plant on the Pennypack Creek provided drinking water for his factory (although, as previously stated, he used Delaware River water for the steel-making process). More important, he was able to sell water to the residents of Tacony. The profitability of these utilities gave Henry Disston & Sons extra capital. In 1887 the water rents produced a cash profit of $4,994.01. In 1922 the Tacony Water Works, by then a worn-out, inefficient plant, was sold to the city of Philadelphia for nearly $1 million.[5]

A second utility was fuel gas, then the main provider of light for urban homes. How Henry Disston & Sons gained control of the fuel gas in Tacony is told in the minutes of the Tacony Fuel Gas Company for 1888–1896, which outline and record the Disston family's financial support for the establishment of utilities in the community. By so doing, the company received water and gas service at minimal fees, the community was guaranteed modern urban living conditions, and skilled labor was enticed to move to Tacony at little expense to the saw works.[6]

A small fuel-gas company existed in the river community of Tacony before Disston's arrival in 1872, but it had neither the finances nor the equipment to provide gas service to the new residents. In 1888 a group of men met in the house of Thomas South to form the Tacony Fuel Gas Company. Five directors were elected: Thomas South, chairman; S. T. Williams; Jon Matlack; Samuel Lambert; and George Fodell. Hamilton Disston was treasurer. They invested money in the company as follows: Disston, $75,000, South $8,000, and the remainder of those present, $8,100. The money was used to purchase the old fuel plant and to provide the pipelines and extensions necessary to pipe gas into the new residential section of the town. To complete the work, the board of directors authorized issuing bonds worth $100,000, repayable by the year 1918. A mortgage on the plant, the lot on which it stood, engines, boilers, pipes, generators, and machinery was held in trust by the Guarantee Trust & Safe Deposit Company of Philadelphia as collateral. By February 1889 the company's stocks were almost entirely in the hands of the Disston family, apportioned in the following manner: Hamilton Disston, 560 shares; Horace C. Disston, 360; Jacob S. Disston, 290; William Disston, 240; and Thomas South, 100. Four others held the 70 remaining shares. A year later, the board of directors read like a roster of Disston company officers: Hamilton, Jacob S., and Horace C. Disston, along with Thomas South and William Miller.[7]

The company prospered, and by 1897 it was ready to pay off all bonds purchased to finance the expansion of the firm into Tacony. At this juncture the gas business was put up for sale. Francis H. Banks offered to buy the company, but his bid was refused. The Disstons themselves were interested in owning the works. The Tacony Fuel Gas Company owed $8,000 to the Disston Saw Works, a sum that had to be returned at purchase. Fearing that a private group might raise gas prices, Jacob S.

Disston offered to purchase the company for $10,500 in June 1897. Of that sum, $2,500 would pay off back taxes owed the state, and $8,000 was to be used to cancel the Disston debt. In July the deal was finalized, and Henry Disston & Sons now officially controlled the gas used in the town. This gave Disston ownership over the community's basic utilities, water and gas.[8]

Electricity was another story. Local Tacony builder Peter E. Costello organized the Suburban Light Company in 1891. Few if any Tacony homes had electricity before 1900. Inventor and futurist Frank Shuman wired his house for electricity when it was constructed in 1896, even though almost everyone else in the town doubted that electricity would be in general use in their lifetimes. Disston & Sons built elaborate steam pipes that ran from the shop engines to a generator, producing plant electricity for the first time in 1911. In the early period, Disston & Sons sold extra electricity to the community.[9]

The houses along Longshore Street, then the main business district of Tacony, were the first in town to receive electricity from Suburban Power & Light. Bob Bachman, born in Tacony in 1906 and a lifelong Disston employee, remembers helping his father install the wires in their own house in 1914. His father purchased the wire at Joe Smith's hardware store and installed the power line by taking up the floorboards himself. The family of grocer Edward Darreff had their house wired for electricity the same year, but kept the original gas fixture because Edward's father was afraid the electric lights would not work. This fear was shared by the builders of Tacony, who continued to build homes with gaslight lines until the 1930s, indicating that electricity came to stay in this suburban industrial village only at a late date.[10]

These company records make it clear that the Disston Company was growing financially with each year. Much of the family's wealth came from land speculation in Atlantic City and Tacony, with added cash flow being generated by ownership of utility companies. But Disston's management style was the key factor in the profitability of the saw business.

The life of Henry Disston allows the historian a unique insight into the work and character of one of the most successful industrialists in nineteenth-century America. His story gave credence to the popular Horatio Alger motif. Here was proof that anyone in America could be a success if he worked hard and persevered. But while Disston exhibited these traits, others exhibited similar qualities without the same success. We must look beyond these superficial resemblances to find an explanation for his unique success. No simple formula provides the key. As we have seen, Disston was benevolent to his workers. He emphasized quality and had the instinct to discover and create new machinery for producing greater quantities of saws. But none of the improvements he made was without risk.

The introduction of laborsaving machinery in some Philadelphia manufacturing firms led to violence on the part of workers who feared losing their jobs. This was especially true of the steel industry when steam hammers were introduced to replace the physical hammering of billets. As dangerous and unhealthful as the old method of hand grinding might appear to us today, the saw-grinders of Disston's day saw worse consequences to themselves from newly designed machines.

In order to allay fears, Henry Disston and his sons personally demonstrated the new machines they introduced to the factory. They showed it was easier for a man to direct a steam hammer than to use a hammer manually; that it was easier to control grinding with a machine than by hand; and that saw handles could be more easily cut with a bandsaw machine. Because the Disstons were a familiar presence in the factory, such appearances to demonstrate new machines were not greeted with suspicion by the workers.

But personal demonstrations did not dispel the fears that machinery would somehow replace factory workers. For this an educational program was begun, to demonstrate that machinery was actually responsible for more workers and higher wages. The company kept accurate records of wages paid to show how machinery not only lowered prices for customers (Table 3) but also increased the workers' take-home pay (Table 4). But the figures mean little until they are compared with wages in other skilled occupations for the years in question. Earnings for skilled laborers outside of the factory were actually declining during the time Disston workers were receiving wage increases. In 1872 the average wage of skilled laborers was $2.64, in 1877 it dropped to $2.18, and in 1880 there was a slight increase to $2.26. These averages, when compared with those in Table 4, show the gains in real income by the Disston workers

Table 3. Price of Disston No. 7 Saw, 26 Teeth, 1870–1889

1870	$18.47/doz.
1875	14.96
1880	13.32
1885	12.31
1889	11.55

SOURCE: William Dunlap Disston, Henry W. Disston, and William Smith, "The Disston History," comp. Elizabeth B. Satterthwaite (1920), 1:76.

Table 4. Wages at Disston in Dollars per Day, 1872–1887

	1872	1877	1882	1887
Long-saw smither	2.00	2.25	2.50	2.66
Hardener	2.00	2.25	2.50	2.66
Circular-saw grinder	2.00	2.25	2.25	2.50
Setter & sharpener	2.00	2.00	2.50	2.75

SOURCE: William Dunlap Disston, Henry W. Disston, and William Smith, "The Disston History," comp. Elizabeth B. Satterthwaite (1920), 1:75.

in the years between 1872 and 1880. By 1880, workers at Disston, despite starting at lower wages than skilled laborers outside the factory in 1872, were making more per hour. Increased production caused by increased use of machinery seemed to be the driving force behind these wage shifts.[11] Machines were clearly valuable to the worker and customer, both lowering the price of saws and increasing wages for the workers. The company also used this information in marketing saws to lumber and hardware men.

As important as Disston's technological visions were, they would not have been successful in the long run without skillful management of the business. Management—the ability to control the flow of capital, administer factory activities, and promote steady work—was the crux upon which firms succeeded or failed, and Disston's business management techniques helped Henry Disston & Sons become America's number-one saw company. Henry Disston's workaholic habits and personality were not the only factors responsible for the company's growth, but his ability to share leadership responsibilities, and his belief in high quality, service, and dependability, were the guiding elements in the company's success.[12]

The day-to-day management style of nineteenth-century industrialists has not received much attention from historians, but enough information about Henry Disston's business life exists to permit some basic observations about his business style. His business letter book for July 3 to August 31, 1872, survives, preserving more than 500 letters dealing with daily business decisions at the Disston plant. "Tacony Real Estate Sales Book No. 1" describes every property sold by Disston in Tacony from 1872 to 1898. The Atlantic City Real Estate Book lists investments in that city and all purchases of equipment at the Disston-owned mill there for the years 1867–1869. Private account books, commencing in 1867, document the company's profits and family dividends. When viewed together, the records reveal how the Disston wealth was accumulated, what was done with the profits, and the company's relations with the workers and with hardware men throughout the nation. These sources yield insights that go beyond the value of hard work and ingenuity.[13]

The business letters written in July and August 1872 give a picture of how the company was run on a day-by-day basis. An average of twelve letters a day were sent out by the firm, usually by one clerk assigned to that task. Only occasionally did Henry Disston himself write a letter, and then the purpose was usually to collect money or reply to a concerned customer. Nevertheless, the contents of all the correspondence emanating from the company, regardless of authorship, makes it clear that either Henry Disston or his half brother Samuel was being consulted daily for responses.[14]

The letters reveal a number of important facts about saw-making and the steel business of the time, and they explain Disston's ability to dominate the saw market in the United States. Saw production was not standardized throughout the country, and saw teeth varied from manufacturer to manufacturer, but Disston understood that teeth had to be patterned according to the wood to be cut. This understanding was behind the company's willingness to manufacture saws according to any request. Henry Disston was more aware than his competitors that saw superiority came from the shape of the saw and the design of the teeth. In the late 1860s, Disston produced inserted-tooth saws. Later he derived cross-cut saws with raked teeth, which made them by far the most effective saws used in America's lumber camps.[15]

Henry Disston's desire to keep up with the latest developments in manufacturing is illustrated by his collection of patent and design drawings of saws from 1835 to 1875, now preserved at the Eleutherian Mills Historical Library in Greenwood, Delaware. These saw designs, as previously mentioned, were tested at the Tacony site before the new factory was built in 1872–1873. Disston's organizational ability and thoroughness are evident in these manuscripts. He methodically collected patents on saws, analyzing and commenting on each. He learned which types of saw blades cut various wood products best. He knew the names of the patent holders and would contact them for permission when special saw orders were received for a specific type of product. In all, his records indicate well over one hundred different types of saw blades conceived during this forty-year time span—a tribute to what might be called today an effective research and development department.[16]

The manufacture of circular saws also required knowledge beyond the design of the teeth. A thorough understanding of lumber-mill operations in general and the precise positioning of bolt holes in particular was required to fill requests for circular saws. This meant that Disston had to be in regular touch with engine-makers and lumber mills and have access to eclectic, generalized, and timely information in order to keep his saws usable. Stories in *The Disston Crucible: A Magazine for the Millman* repeatedly refer to mill frames and mill equipment. A letter of 1872 explains to a customer: "The shingle saw spoken of . . . would not be worth much more than scrap steel to us for the reason that we would not be able to sell it except to some party using the same machine and I might not receive an order for same for years." The size and location of tug pinholes determined whether a saw could be used by an engine of a specific make. Disston eventually set the standard for the industry, giving those who purchased saws the assurance that he could fit each of the lumber-mill frames then in use.[17]

Basic to the Disston company's success in capturing the American saw market was the ability to respond promptly to special requests from mill owners for specific saws. Disston began stockpiling the most frequently used saws so they could be delivered sooner than saws that had to be made up after the order was received. This encouraged mill owners to purchase machines that used the most readily available replacement blades. The scale of Disston's business thus gave the company considerable influence over the machinery being used in the lumber companies.[18]

An incentive policy of the Disston company also fostered the purchase of Disston saws. Disston accepted "old" saws as trade-ins on new saws, but only if they were "of our make and would make a saw not below 48 inches in size." This offer, combined with a 20 percent discount for steady customers, made Disston saws exceptionally attractive.[19]

Disston underscored this rebate policy with liberal terms for repairing saws sold by his factory. The saw-reconditioning portion of his business soon rivaled the manufacture of new saws in profitability. Resharpening and retensioning went on throughout the life of the firm, but it was during the early days of the firm that its importance as a money-maker was discovered. About 50 percent of the correspondence in the 1872 business book is related directly to these repair services. If the saw could not be repaired, it could be exchanged for a new one with a discount of up to 50 percent—if it was a Disston saw. This could eliminate untimely delays in saw replacement and shipment. Disston's reputation for honesty and integrity fostered such exchanges.[20]

Dealing with Henry Disston guaranteed the purchaser the most up-to-date saw utilizing the most advanced steel technology. Disston's advice on technical matters was honest and to the point: "We have never known that type of steel to be cut square on the ends," he wrote. "Our boss melter has melted hundreds of tons of it in Sheffield [England] but never knew it to be cut square." He rarely disappointed a customer who requested information about where a particular product could be purchased. The company responded, on average, within three days after receipt of a letter, making the firm a favorite with hardware men who needed quick responses for their customers.[21]

If Disston could not answer a customer's question, he referred to someone who could. Thus, he wrote to Peabody Saw of San Francisco that "Mr. Baxter D. Whitney of Winchester, Mass. makes a specialty of this article and we are informed, those of his manufacture are much sought after and give entire satisfaction."[22]

Collecting money was difficult for a firm like Disston with business accounts all over the nation. The best method of dealing with the issue of late payment was prevention. Letters in the business book indicate that it was common to take cash in advance or to check carefully into the background of hardware or lumber men who wanted to open an account. Some letters requested confidential information from firms Disston knew well and with which he had a working relationship. Despite his nationwide business, Disston seldom was left without payment for his products.[23]

Once an account was opened, a second phase of his collection technique commenced. Henry Disston was inflexible when it came to accounts due. Bills were to be paid promptly according to contract. He gave hardware and lumber men thirty days to pay, after which a letter was sent. Written by a clerk at Henry Disston & Sons, a typical letter read:

> We have heard nothing from you as to the arranging of your account in answer to ours 15 inst., and as a consequence according to the terms of the agreement existing between us your account is closed. [I] cannot ship you any more goods upon orders received until you arrange what is now overdue. We regret being obliged to pursue this course and trust you will make arrangements at once to have the embargo taken off so that we can go on with our business relations heretofore. Trust you will receive this in a kindly spirit and respond at once.[24]

When faced with a more serious financial matter, Henry Disston wrote the letter himself and was far more forceful. In a dispute over the commission paid to a hardware man, Disston wrote:

> I found your merchant . . . to do . . . more harm than good to us. I told you in a former letter I gave him $100. for what I could get a boy to do for 5 or 6 dollars. I could have done just as well if not better without him and when he was paid for his services I took

the precaution of settling with him in the presence of witness Mr. Rosencrantz.

In another case Disston went right to the point: "That sucker owes us $7,000. . . . It would be best to get judgment against that man anyway. I gave [master smelter Jonathan] Marsden $600 to have that case prosecuted."[25]

Another area of great concern for Disston was patent infringement by other companies. In the period between 1850 and 1875 he brought five cases to court, and once Disston himself was taken to court. The most unusual case occurred when Disston traveled to Toledo, Ohio, in the spring of 1874 to look at a saw-grinding machine that he suspected was in violation of a Disston company patent. Expressing his admiration, Henry asked the owner David M. Mefford to bring his machine to Philadelphia for use in his factory. A deal was struck, and Mefford came to Disston Saw, set his machine up, and began working. The next day, an astonished Mefford was served with a patent fraud complaint signed by Henry Disston. Disston not only had Mefford accused of patent violation in Disston's own judicial district, but also had the machine as proof. Although Disston was unsuccessful in getting a conviction, he had made his point: Don't copy Disston patents or you run the risk of court action and inconvenience.[26]

The Mefford incident illustrates the lengths that Henry Disston would go to protect the notion of a fair marketplace where just competition ruled the day and where success was determined by hard work and a good business sense. Mefford had violated this creed. Further, Mefford had broken the businessman's code of following fair rules in the marketplace. Henry Disston and others of his standing needed such rules if they were to succeed in stabilizing their businesses, and Disston proved to be a forceful and strong-willed businessman when his company was threatened by those who lacked ethics. Frustration because of the lack of convictions in patent fraud cases eventually led Henry Disston to adopt a company policy of secrecy in discussions of all newly developed products, a practice that was passed on to other family members. Breakthroughs in steel-making processes, methods of hardening steel, and development of new products all became carefully kept company secrets.[27]

The most common letters in the 1872 correspondence dealt with requests for price lists and the *Disston Lumberman's Handbook,* a special catalog of saws and prices for lumbermen. Information about tensioning and repairing saws, although of little practical use to buyers, was found throughout the handbook. Even if not requested, price lists were automatically included with responses to inquiries about product information. Disston felt that price lists were worthless until they were in the hands of a prospective purchaser.[28]

Other letters concerned saw repair or exchange, and information about saw-making requirements. Disston always answered courteously, even when the problem was not the company's fault. In response to a complaint about a saw that did not cut properly, Disston wrote the owner, "Your saw was made precisely as ordered. . . . It is more than likely that it has been thrown over in running. Such is the case at times. If you return it we can change it by hammering." Another such letter received similar consideration. "We can repair your two circular saws by hammering and make them as good as they ever were which would be much better than any trade we could make especially for you. . . . Write us when you send the saws just what your trouble is." For burned saws Disston made no promises:

> We do not guarantee any saw that has been burned as there always is great danger that all the nature of the steel has been burned out. We do sometimes succeed in making a good saw out of a burned plate, but it is a rare thing when the plate is badly burned. We will repair the saw you now have that is burned, but will only do so at your risk. The price for repairing burned saws is 2/3 [the] price of a new saw of same size.[29]

Shipping mishaps, saws failing to arrive at their appointed destination, and complaints about workmanship were the subjects of a smaller segment of the company's correspondence. Typical of these letters is one written to the Smithfield Fulton Company of Pennsylvania: "Our shippers had made a blunder and sent your saw to Springfield, Ill. We would write at once to the Exp. Co. there." Slow delivery sometimes held up supplies, as was the case when Disston was unable to forward bandsaws because of a lack of steel. The matter was cleared up at last: "This day [we] received the right kind of band saw steel [the first] since commencement of the trouble in Prussia." A quick note to a hardware man reassured him: "We have again commenced to receive our regular supply of French Band Saw Steel and have some of the sizes you want."[30] Sometimes Disston found it necessary to renege on warranties because customers had damaged the product:

> The mill saw returned to us, it has been examined and we really cannot see that wear [*sic*] at fault in any particular. We observe they have cut a large piece out of each end we say cut for we do not know how it could have been broken by fair means, as the saw

> is if anything a little mild in temper rather than hard and he has punched large holes in the ends thereby rendering it entirely worthless to us except as scrap. If the party had returned it in any kind of decent shape we would willingly have exchanged it for anything he might have wanted. We do not consider it returnable as it is. If he will tell us why the saws will not work satisfactory we may be able to remedy the trouble for him.[31]

Requests involving special projects usually brought advice on the best method for the undertaking:

> The article in question would be difficult to make solid and would cost as much as if the plate were left without cutting out the center. We would suggest that you have this circle made in segments of say 1 to 10 pieces which would cost much less than to have the plate solid. We of course do not know what you wish to do with the article but are of the opinion that segments would answer. We could not give the cost until we were advised more fully.[32]

When questioned about the use of inserted toothed saws, Disston's gave a full description that included specific information about the saw's cutting properties:

> We have to say that they are in pretty general use and are used for cutting all kinds of lumber. And if kept in proper order will cut fine and smooth and certainly has no more tendency to chop or split than another tooth. It is true that the harder and more solid the wood to be cut and the greater speed and feed used the more teeth you would require. But in a fifty four (54) inch Diameter Saw we are putting in twenty-six (26) teeth which would be about six (6) inch space and with this number of teeth we cut all the general kinds of lumber and do the work well.

Buying in quantity could mean a lower per-unit cost to the purchaser, as proved to be the case when a clerk advised a manufacturer to say "we can furnish you the spring as per sample in qualities of one thousand (1000) at three (3) cents each You to pay for dies for bending which will cost $12.00."[33]

Inquiries sometimes prompted extensive investigation by Disston and occasionally resulted in a new product: "As to the McKenzie Square we have to say that McKenzie is dead but we have been unable to find his

1. Henry Disston (1819–1878), founder of the Keystone Saw Works, later incorporated as Henry Disston & Sons. Disston came to the United States in 1833 at the age of fourteen. He was apprenticed to a Philadelphia saw-maker, before launching his own business in 1840. After a series of successes as well as setbacks, Disston's company grew dramatically, so that by 1859 it employed more workers than any of its local competitors. By the end of the Civil War, Disston had captured a share of the national saw market. ("Disston History," vol. 1)

2. Henry Disston & Sons industrial complex in Tacony, 1943. From 1872 to 1899, the operation was gradually moved from its Laurel Street location in Kensington to Tacony, on the outskirts of Philadelphia a little south of where the Pennypack Creek feeds into the Delaware River. This aerial photograph shows the plant during its most productive years, when Disston's company was the world's largest manufacturer of saws. (Tacony Public Library)

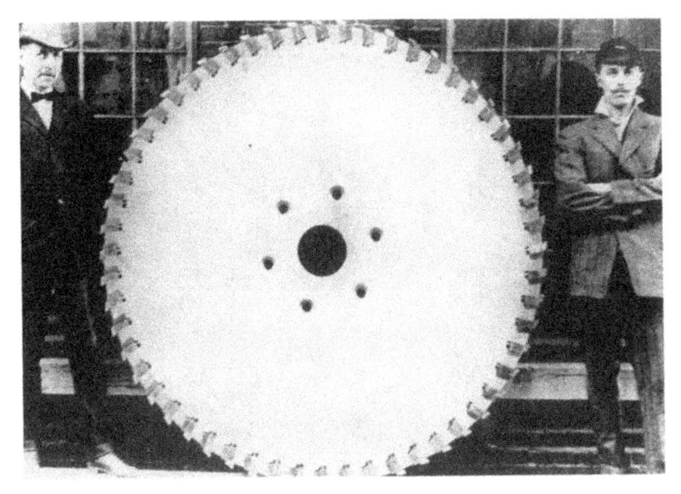

3. William D. Disston, grandson of Henry Disston, standing to the right of an inserted teeth circular saw in 1910, during his apprenticeship. A shop foreman stands to the left. Henry Disston required all his sons to apprentice in the shops before taking on a managerial role. William became a partner in the firm in 1878, the year of his father's death. His son William D. (shown above) became operating manager and CEO during the Great Depression. (Private Collection of William L. Disston)

4. Disston managers from Laurel Street plant, 1881. In the front row, left to right, are Disston family members Horace (fourth from left), Hamilton (fifth from left), Albert (sixth from left), William (seventh from left), and Samuel (tenth from left). Hamilton had become president of the company in 1878. (*Disston "Bits"*)

5. Steelworkers with crucible steel pots and tools, 1884. Henry Disston and his sons used family connections in England to find skilled workers and to keep up with new steel-making technology. English smelter Jonathan Marsden brought the crucible steel process to the company in the 1850s. In the days before chemical testing, these skilled craftsmen were modern-day alchemists. (*Disston "Bits"*)

6. Disston steel plant, built in 1900. The crucible steel process gave way to the Bessemer process early in the twentieth century. Not until the 1930s was hot-metal testing introduced in the Disston plant. Here, iron ingots are stored in the yard ready for melting. (Atwater Kent Museum)

7. Crew of No. 7 Crucible Steel Melting Furnace, 1919, wearing protective glasses, clothing, and footwear. Steelworkers were forced to adapt in rather primitive ways to the dangers of the crucible furnace. Work in these outfits was particularly difficult in the hot summer months. However, the Disston plant, given the nature of the work, had an enviable health and safety record. (*Disston "Bits"*)

8. Steel plate cutting machine. Here steel was cut to the appropriate saw size and then sent for teething. By the 1880s, machines had replaced many of the old hand processes. ("Disston History," vol. 2)

9. Circular-saw teeth-cutting machines. Saws were teethed while the steel was still soft. Next, the saws had to be tempered in a special hardening bath. (Tacony Public Library)

10. Smithers balancing circular saws, 1880. Smithing, a highly skilled trade that took years to master, involved the hammering and blocking of saw blades until perfectly flat and smooth. Here smithers work by windows to check for light, which indicates high or low spots in the steel. Holders patiently wait to hold saws on the anvils. (Tacony Public Library)

11. Disston saw smithers and holders at Disston's Laurel Street site, 1869. Smithers are holding long pieces of steel used to evaluate the flatness of the steel surface. Both holders and smithers have canvas protective flaps over their aprons to protect them from the sharp edge of the blade. (Atwater Kent Museum)

12. Rows of large circular-saw grinding machines. Grinding was needed to taper blade thickness. Using these machines, a worker could finish six saws for every one using the old hand methods. However, many of those who operated these machines would die of silicosis, a fatal disease caused by the dust created by the natural sandstones. Disston Precision Inc. still has one of these machines, but it is not functioning. ("Disston History," vol. 1)

13. Master smither William Rowen. Like those before him, Rowen learned his trade from oldtime expert smithers at the Disston company. Here he uses a flat steel tool to see bumps and indentations in a circular saw. He had twenty different hammers to straighten and tension saws. Once Rowen became foreman in 1939, he inspected every circular saw that left the plant. (Atwater Kent Museum)

14. File shop, prior to 1920. Here files were made for sharpening saws. The straps connected each machine to a steam-driven engine. ("Disston History," vol. 1)

15. Worn-down grindstones from the Disston plant, 1912. No longer useful in the factory, these stones are to be used as building stones for the Tacony Baptist Church on Disston Street. Artificial grinding stones were not introduced into the plant until the 1930s. (Tacony Public Library)

16. Handsaw grinding department, prior to 1920. Handsaws were produced separately from the circular saws. Disston & Sons produced every imaginable sort of blade for cutting wood or metal. ("Disston History," vol. 1)

17. Handsaw sharpening department, prior to 1920. Once the teeth have been cut and the blades hammered, rows of workers use files to sharpen saws one tooth at a time. ("Disston History," vol. 1)

18. Bandsaw smithing department, prior to 1920. In 1865 Henry Disston brought the first bandsaws to the plant, where they were used to cut saw handles. The plant therefore produced bandsaws for use in the plant as well as for sale. Here large bandsaws are pulled along specially built tables and smithed one section at a time. ("Disston History," vol. 1)

19. Smithing handsaws, pre-1920. A last tension check is being made before drilling holes for handles. ("Disston History," vol. 1)

20. Handle-cutting shop. Using bandsaws, workers cut saw handles from three-year aged applewood. This picture was taken before 1920, when steam-driven engines powered every bandsaw. ("Disston History," vol. 1)

21. Shellacking saw handles prior to 1920. This was one department reserved for women employees. While men and women were hired as factory labor, they rarely worked side by side except in the office. ("Disston History," vol. 1)

22. The Shipping Department, where orders for saws were filled for delivery all over the world. ("Disston History," vol. 1)

23. Disston office workers in 1918. This is a surprise party for one of the men before he left to fight in World War I. (*Disston "Bits"*)

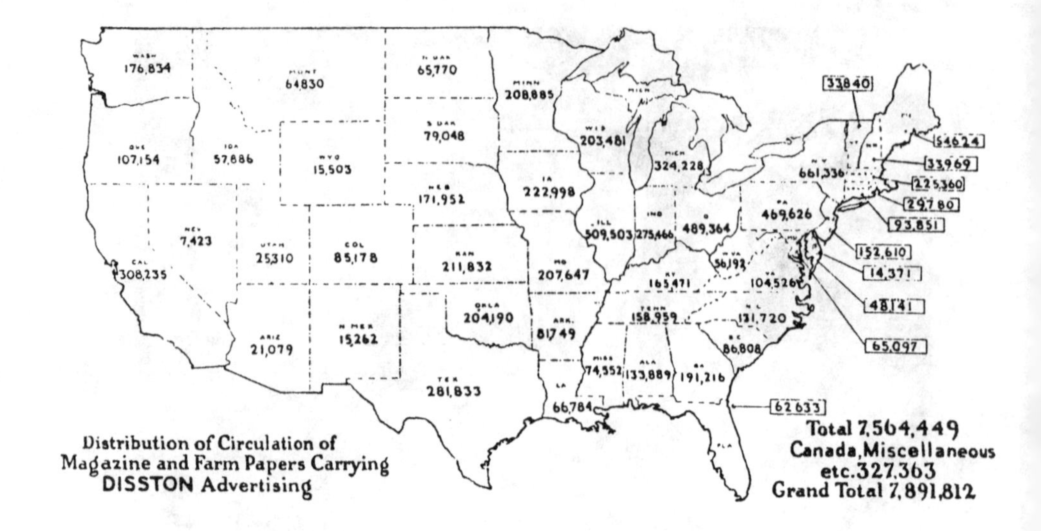

Telling the United States

The figures in the different states represent the number of copies of magazines in which Disston advertising appears.

We are telling this tremendous number of people why they should use Disston Saws and Tools. Notice the number in your state. Figure the direct influence on the section of the community you serve.

Many of your customers must see this advertising. Many who should be your customers will see it. We can convince some of their need for Disston Saws and Tools.

But your profit lies in your *selling* not our *telling*. The latter helps wonderfully, but you must do your part.

Let your customers know, either by window or counter displays, or by word of mouth that you sell Disston Saws and Tools.

If you do we will consider our outlay for advertising well spent.

HENRY DISSTON & SONS, Inc.

Keystone Saw, Tool, Steel and File Works

PHILADELPHIA, U. S. A.

Disston advertising appears in the Saturday Evening Post, Literary Digest, Popular Science Monthly, Scientific American, Popular Mechanics, and in twenty-nine leading farm publications reaching every part of every state from coast to coast. The total circulation of these publications is approximately 8,000,000 copies each issue. Far more than that number of people read them.

24. Map from *Philadelphia Made Hardware* catalog in 1920, showing that Henry Disston & Sons had an active advertising campaign in every state of the union. The company advertised in magazines as diverse as *The Saturday Evening Post* and *Popular Mechanics.*

25. Disston industrial display, 1894, when Henry Disston & Sons was the world's leading sawmaker. Elaborate in size and style, Disston's displays used real saws and a world globe to depict to viewers that Disston had indeed captured the world saw market. ("Disston History," vol. 1)

26. Henry Disston & Sons of Wisconsin, 1955. Although no written records exist in company files, this picture is proof of Disston management's plan to open a plant in Wisconsin to produce a special gasoline engine for a one-man chain saw. New management decided that Disston had to enter the power chain saw market or lose its competitive edge in the saw business. Among those standing are: Guy Conrad (fourth from left), William L. Disston (sixth from left), John Thompson (ninth from left), and Jacob Disston Jr. (tenth from left). Walter Gebbhart is kneeling, second from left. (Private Collection of William L. Disston)

27. Disston Precision Company, 1991. Disston & Sons was bought by H. K. Porter Company of Pittsburgh in 1955. Porter broke up the various Disston divisions and sold them in the 1970s. The three factory buildings shown here are all that is left of the mighty Disston factory complex in Tacony. The Disston Precision Company still uses some of the original teeth-cutting machines to produce a special circular saw to cut hot metals. (Atwater Kent Museum)

widow, having scoured the whole eastern country but no tidings. We intend getting up something in the square line that will take the place of the McKenzie and will advise you when done." Disston's willingness to generate a new product to meet a customer's need was just good business. However, when Disston could find an existing manufacturer of the product requested, it was his practice to refer the customer to that firm for service: "We manufacture the electric x cut saws for the Schweitzer Mfg. Co. New York who control the prices. We have sent your letter to them with the request that they will furnish you best prices as desired."[34]

Disston understood steel and saws down to the most minute detail. "We have no 48″ saw as described on hand consequently cannot send at once as desired. We will try to get you one in one week if that would answer. That would be the best we could possibly do." Promptness of delivery was vital to the customer. A typical letter answering such concerns read: "We will forward the 50 inch saw at once in about two weeks." Knowledge of saws caused Disston to question imprecise orders: "Your favor of 22nd inst. recd. but you have not given us countersink size, which we must have before we go on with your order—also state whether saw runs horizontally or perpendicular," or "In reply you have not given us the length wanted. We are sending out various lengths for different machines so you will see the importance of giving length." Another letter informed the customer: "We do not know just what you mean by 'saw sheets' we do not manufacture any thing we call by that name, inform us what its used for, with a description or sketch perhaps we can [*sic*]." Ott-Hill & Company in Wheeling, West Virginia, was informed politely that it could advantageously consider changing its order: "We do not manufacture perforated saws not deeming them by any means as good as our solid plate saws are." Advice on how to handle steel problems was also dispensed freely to customers, as when Disston wrote: "Dip the cutter when hot in salt and cool in cold water or you will get them too hard."[35]

Henry Disston's management style was far from autocratic. He put great faith in the capacity of those around him to make decisions. He delegated authority well, and he seldom interfered with a department unless the quality of the product was poor. Whether the issue was the making of price lists or his half brother Samuel's decision to give a reduced price while visiting a client in another city, the letter book shows Disston was at ease and supportive regarding his employees' initiatives and input.[36]

Thus, Samuel Disston (1839–1908) had as much to do with the company's success as did Henry. Born in Nottingham, England, in 1839, Samuel was educated in the schools of Nottingham and as an office boy at Disston Saw Works. Hired by Henry in 1850 when he was 12 years old, Samuel served the firm continuously. He was a devoted Republican, a member of the Union League, a Philadelphia school board member, and president of the Columbia Club. He died on June 27, 1908.[37]

The letters for a two-month period abound in evidence of Samuel's competence, energy, and enthusiasm on behalf of the company. The charismatic and tireless Samuel regularly toured the United States, building up a network of hardware distributors committed to selling Disston saws and tools. One hardware distributor remembered forty years later that he had met Samuel in his father's store. He recalled vividly his father's enthusiasm, excitement, and respect for Samuel. Little wonder that letters from hardware distributors and lumbermen requesting meetings were referred to Samuel Disston. One such letter brought this response to a New York City customer: "Our Mr. Samuel Disston will be in New York in a few days and will call to see you in regard to circular saws." Samuel was a master of the personal touch, and Henry understood its value in cementing longtime business contacts. A lesser manufacturer with a less solid ego might have had second thoughts about this charismatic operator and his free and open business style, but it was Samuel, who lived into the twentieth century, who finally provided the leadership and personality to allow the company to dominate the saw industry.[38]

Foreign saw sales in America were another issue that Disston confronted directly. His belief was that one should not try to undersell foreign saw-makers and placed his faith instead in the eventual endurance and triumph of a superior product. At the start, Disston's export trade was handled through New York commission houses; he then branched out. Having discovered that the German and English competition dispatched personal representatives, he patterned his approach on theirs. Thereafter, once every two years Disston would send a man around the world on an eighteen-month selling tour. Eventually Disston realized that direct company representation yielded far more profit than employing intermediary agents. From these beginnings, Disston developed a system of separate business houses in major urban centers around the nation.[39]

Disston promoted autonomy and entrepreneurship among his workforce as well. Each factory and shop in the plant's loose network contained within its walls inventors and risk-takers who were encouraged to improve quality and production within their specific department. Most notable in this capacity was the work of Thomas Disston, who in partnership with Henry owned the jobbing shop. Thomas had twenty-seven patents, far more than Henry. Also important was the freedom given skilled workers like smelter Jonathan Marsden, who was sent to England each year to hire workers. The ability of those men to make independent decisions, and Henry's latitude in that area, added to the firm's success.[40]

Disston realized early that production depended on people. His basic tenet was respect for the individual. During his daily visits to the shop, he treated foreman and worker alike with compassion and respect, and he was the prime motivator of production.[41] He was at the plant every day overseeing the business. Disston always followed his own advice and stayed close to his knitting, which in his case revolved around knowledge of steel and saws. His commitment to improved production methods is shown by his purchase of the steam engines used at the Centennial Exposition in Fairmount Park for his new Tacony plant when he was ill with the first stroke. The engines were distributed around the plant and used a line shaft and belt-driver system to power as many as fifty machines per shop. Many of these machines were still being used in the plant as late as 1950.[42]

Yet the basic philosophy of the organization put more emphasis on achievements than on new machinery. Disston practiced hands-on "value management" practices—that is, the cultural foundation of the business was the common values of quality and integrity, with the staunch ethic of Henry Disston himself involved in each step of the process. Complaints about quality disturbed him deeply, especially if there was any validity to them. After learning about a lapse in workmanship, he wrote to one customer, regretting "that we have to hear complaints like this. I write you it shall not occur again if I have to send every man away from the factory." When Henry Disston investigated a similar complaint and found that the foreman was not using the care he should, he banished him from the department and wrote to the customer: "I can't understand why workers can't make the product right the first time." But such justifiable complaints were the subject of only two letters in a two-month period; the vast bulk of the correspondence reflected customer satisfaction. The workers were made well aware of Disston's convictions and his penchant for perfection.[43]

When courting a new business relationship, Disston would boast that he had the best workers and technology in the business:

> We have just had a young man here by the name of Boetticher of your city engaged in the hardware trade. We talked with him about circular and other mill saws, but he informs us that you do the principal part of that trade. We have been anxious for some years to supply the trade in your section with goods in above line.

> And knowing that we can compete both in quality and price with any house in this or any other country. We would very much like you to give us a trial and if we do not excel or give satisfaction we ask no pay. We have proven the matter to thousands of men in this country, to which any hardware man can testify as to our short saws. But now we wish to let you know the difference between our saws and those of other makers. We will be much pleased to receive a trial order. You shall not lose by so doing if you do not gain but we know you will gain. We have the best machinery and the best workmen the world can produce, and will warrant all goods to be the best quality and manufacture.[44]

Disston had streamlined his production lines with new saw-making equipment and developed a system of warehousing popular saws and blades with far more efficiency than his competitors. His commitment to quality work and service gave his customers a great advantage in their operations. Spurred on by improvements in transportation and expansion in the housing market after the Civil War, Disston's business grew rapidly. Skilled workers, especially steel smelters and mechanics, could not be found quickly enough to fill new orders for saws. Eventually, workers were imported from England to fill the gap, but still some delays were inevitable:

> We admit that we were behind time in forwarding this saw but it was unavoidable on our part, we did the very best we could. We have been so much behind our orders on mill work for the past season that it has been impossible at all times, owing to circumstances over which we have had no control to fill orders at the time named when orders were received. But we are now making arrangements [hiring more skilled workers and warehousing popular saws] so that we will be able to fill orders more promptly in future. We regret the delay, and trust your customer has not been seriously inconvenienced and that the saw when received will prove satisfactory.

Some requests were more easily filled:

> Your favor of 25th inst. recd. In reply say the 50″ saw referred to will be shipped on Friday 25 inst. We regret not being able to fill this order more promptly, but we are so much behind orders that it has been impossible to forward sooner.[45]

Throughout his career, Henry Disston had a bias for action. His invention of the skewed-back saw is one example of how he put ideas into action. His trip to Paris and the subsequent discovery of a workable bandsaw is another. In his management style this trait was manifested in impatience with those around him who did not perceive and seize opportunities:

> We sent you word that we would give you the job at from 18 to 21 rollers of steel per week [a month ago] . . . you not coming in answer when we offered the above amount left us with unfilled orders. We made other arrangements and therefore do not need your services.[46]

Another impetus to Disston's business success came from his grasp and utilization of the power of advertising. Henry Disston was in the forefront of catalog and newspaper advertising. His use of pictures with products, particularly for international markets, showed both imagination and foresight. He advertised in newspapers in Mexico and Canada as well as in the United States as early as the 1870s. His products and trademark were pictured, and statements from users attested to the Disston quality. He had agents in every large city, and new offices for direct service to customers were opened all over the continent.[47]

Disston also took every opportunity to be part of local parades and celebrations. To him such events presented an excellent opportunity to get the company's name and business before the public. An attractive, properly decorated float and 100 workers in line demonstrated the significance and importance of a company as much as any advertisement on a page. The practice of participating in parades was continued by his sons after his death.[48]

After Henry Disston's death in 1878, the firm continued to change with the times. When new techniques for selling products began to appear at the beginning of the twentieth century, Henry Disston & Sons was among the trend-setters. As early as 1913, when advertising was in its novice stage and just becoming an important technique for firms that sold in wholesale lots to distributors, Disston & Sons was in the forefront of the movement. Of particular interest to the firm was the hardware industry. Not alone in this concern, Disston formed a confederation with Miller Lock Company, North Brothers Manufacturing Company, Fayette R. Plumb Inc., and the Enterprise Manufacturing Company of Pennsylvania to publish the *Philadelphia Made Hardware* catalog.[49]

The magazine offered free services to all hardware stores. There were free window displays, guaranteed to generate attention and interest and to increase customers and profits. Each issue was designed around a specific topic, with advertisements appropriate to each of the firms, and was provided free (including postage) if the hardware stores used products from at least two of the generating companies. The displays were not always about products; sometimes they were topical. During World War I, when patriotism was in vogue, the magazine pointed out: "This is a time when patriotic expression or displays immediately draw a kindly feeling toward the dealer who brings them to the public notice." The display in question consisted of red, white, and blue "Flying Window Trim" and a 30 × 40 inch poster of Independence Hall. Eagles hovered about, as if to protect that most cherished American landmark. "The combined effect is so original, so delightful, so significant of the spirit of the day, that it's a down right missed opportunity if you don't make use of it."[50]

A second concern was the training of salespersons. Most issues of the magazine devoted space to this problem. Authorities concurred that the qualities most desirable in "green clerks" were good health and natural courtesy, but no store could afford to ignore the training of clerks—a primary lesson being that there isn't a thing in the world like truthful salesmanship. Many hardware men argued against formal training in the belief that it "generated untoward ambitions, on the grounds that this makes [the clerks] dissatisfied and want more money. . . . The modern merchant has no patience with such a theory. . . . It was once thought that one-price and money back being the modern business man's motto, anyone could sell goods on that principle. . . . [But] sales depend largely upon the knowledge, skill, tact, courtesy, and all-around ability of the salesman. . . . The Era of Education has had to begin." For sales staff came the recommendation that wages be paid in proportion to quantity of the sales made—"As you observe, it is a question of vision."[51]

Another article emphasized the importance of female customers in what once had been solely a man's world. The magazine maintained that the goal of advertising must be to bring the customer to the marketplace. Women had great influence over their husbands. Therefore, increased comforts and accommodations for women would encourage them to stay longer in the hardware store. The magazine called on hardware stores to consider installing a women's rest room in the back of their facilities. Picturing one such room, the magazine pointed out that if the room was made commodious enough the women would not object to their husbands' spending more time—and consequently, more money—in the store. The owner of a hardware store in Ontario, Canada, wrote of the idea: "In our opinion no money can be spent in advertising more beneficial than laying aside and equipping a ladies' rest room & toilet."[52]

Still other parts of the magazine emphasized motivational slogans and similar help for hardware men—for example, "Many a good sale is lost because the salesman gets in too big a hurry. Always let the customer take his or her own time." "Good advertiser" contests rewarded particularly clever and enterprising ads for newspapers and magazines. An analysis and critique of the winner's ad alongside the ad itself helped hardware men design more lucrative ads for themselves:

> Unpretentious in size and arrangement, this advertisement embodies strong selling features which are worthy of attention. The ad is designed along the lines that have proven successful in department store work. Note the large variety of subjects it covers, also the display of prices. The few words of copy in each case have been chosen carefully with intention of merely stating the salient points of each tool. Possibly a further elaboration might have been better. Also in the opening paragraph the statement, "we guarantee every tool," should assuredly have been a selling point to all.[53]

A second publication appealed to another audience. Published monthly after the turn of the century, *The Disston Crucible: A Magazine for the Millman* performed a number of important advertising functions for the firm. It was the company's link to the lumber mills of the country, its way of informing customers about plant improvements, explaining saw repair, and imparting current knowledge about wood, saw blades, and sharpening devices to woodsmen. One edition would show the redwoods of California, another the cypresses of Florida. "Who's Who in the Saw World" featured Disston representatives stationed throughout the nation and established them as valid and valued representatives to the saw consumer, feeding the belief that the most knowledgeable and accessible saw sales representatives in the country worked for Disston. The editorial emphasis in the *Crucible* was not on selling techniques but on saws, and specifically on the assertion that Disston was the most knowledgeable, dependable, skilled, and ubiquitous crafter of quality saws in the world. Disston & Sons' production of two magazines reflecting divergent customer strategies is a barometer of the sophistication with which the company approached advertising.[54]

These new selling techniques catapulted Disston far ahead of other firms in America. The advertising campaign stretched to the shores of Australia. Major Leslie Jackson Coombe (1883–1944), making a business trip to Australia in the 1920s, found that "Disston and Atkins have done everything possible to wipe out competitors in this line [saws] and have pretty well succeeded." Superior advertising aside, Coombe added that "English cross-cut saws are not up to Disston standards."[55]

Writings about Henry Disston point out that his penchant for quality, integrity, and accelerated production were fundamental to his success as a manufacturer. Although true, this observation is far too limited in its perspective. Documents describing his actions as manager of the Disston firm tell a different and more multifaceted story. His management genius, his skillful manipulation of the saw market, and his use of up-to-date advertising techniques, which were embraced by his sons, were every bit as intrinsic to his firm's success. Saw reconditioning, rebate offers, discounts for steady customers, and superior knowledge of saw frames and steel helped to market Disston saws. Advertising that featured woodblock pictures of his product, and later company books and magazines, were aimed directly at the company's best customers—lumbermen and hardware distributors. Prompt delivery, a national direct sales force, Samuel Disston's face-to-face contact, prompt correspondence habits, and a firm and consistent credit policy resulted in efficient distribution and payment of bills. It is clear that Disston's success was based as much on his management style as on mechanization and the quality of his steel.

Paradoxically, the financial success that permitted establishment of a paternalistic community also encouraged a larger workforce and a national line of customers—outcomes that ultimately forced a transition away from paternalism.

CHAPTER

4

Pluralism Comes to Tacony, 1885–1918

THE FINANCIAL SUCCESS of Henry Disston & Sons provided the capital to develop Tacony. Disston provided monies for streets, homes, and water and gas works in a time when the urban landscape was not the exclusive responsibility of city governments. Paternalism provided the setting for the distribution and growth of this capital. The success of paternalism in these early years hinged on four conditions present in Henry Disston's Tacony: a small-town setting focused on one industry, the presence of a benevolent manufacturer who valued workers as people, a community removed from the temptations of the city, and a profitable business with excess capital that could be used to benefit the workers beyond the plant gate. After 1900, pluralism changed the equation, limiting the influence of Disston's brand of paternalism and encouraging the company to institute programs in industrial welfare.

Between 1885 and 1918 three growth patterns emerged to bring pluralism to Tacony. First, what had been primarily an English village became a multicultural town. New immigrants included Irish, German, Italians, Jews, and blacks, and ethnic pluralism resulted in religious pluralism. Each group settled in its own geographic area of the community and each established its own church, so ethnically conscious that there were separate German, Irish, and Italian Catholic churches. Second, Tacony moved from being a one-factory community to having more than a half-dozen medium-sized factories located along the Delaware River. Industrial pluralism brought with it competition for workers from within and without the Tacony community. The size of the workforce needed by the factories along the river encouraged the building of the first trolley line into this area of the city. This was the third factor—the advent of a trolley line changed the community's internal structure and increased the labor pool for firms like Disston. No longer did workers have to live in Tacony, making property ownership by Disston less effective as a means of controlling workers. For the first time, residents could easily commute to jobs in Center City Philadelphia, and children could go on to high schools located in adjacent communities. Workers could now live miles from the Tacony plant and not be part of Disston's paternalistic community.

Pluralism spelled the end of Disston's brand of paternalism. Property deeds could not control such a diverse and mobile population. So the company adopted programs of industrial welfare, such as sports teams, fathers' clubs, company picnics, company newspapers, and a cafeteria, as ways of unifying the workforce behind the company. As we shall see in the next chapter, such programs soon supplanted an obsolete and outmoded paternalism.

These changes were put into motion with the death of Henry Disston in 1878. As founder of the firm, Henry had a well-deserved reputation for knowing most of his workers personally and for being a benevolent boss. His heavy investment of company money in the community was well known. After Henry's death, Hamilton Disston became president of the company, but much of Hamilton's time and money would be spent on adventures far removed from Tacony.

Hamilton Disston was born on August 23, 1844, and at the age of 15 became a full-time apprentice in the factory on Laurel Street. When the Civil War began, he was a member of the Northern Liberties volunteer fire company. This promoted a special camaraderie between Hamilton and the workers. As the many fires during the winter of 1861–1862 called Hamilton out of the factory, however, his father became annoyed and ordered the foreman to cashier him if he responded to one more fire call. Hamilton continued to answer fire calls, but the foreman never told Henry and never fired Hamilton. About this time Hamilton twice enlisted in the army. His father on both occasions paid the enlisting bonus and brought Hamilton back to the mill. By this time the charismatic "Ham," as he was popularly called, had developed a strong following among the younger workers in the factory. He and 100 youthful laborers announced that they had formed the Disston Volunteers and enlisted en masse in the army. A frustrated Henry Disston gave his son a reluctant blessing and fully equipped his contingent for war. He even went so far as to pay half again what the government paid soldiers so their families could survive while service was rendered to the nation. When the war ended, the volunteers found their jobs waiting for them. Hamilton returned to the factory as a partner, and in 1865 the firm was renamed Henry Disston & Sons. Horace C. took charge of the steel works in 1875, and William became a partner in 1878. The youngest son, Jacob S., studied business at the University of Pennsylvania and entered the firm in 1882.[1]

With the elimination of the volunteer fire companies in 1870, Hamilton became active in city politics. Like fellow members of the Northern Liberties Hose Company, he supported Republican candidates and helped elect John A. Loughridge, a journeyman in his factory, to the post of prothonotary to the Court of Common Pleas. During the 1870s Ham, along with James McManes, William R. Leeds, and David H. Lane, dominated Republican nominations in the city. Later he was successful in unseating the powerful McManes, the city's gas works czar, when they disagreed about a candidate for receiver of taxes. Although never elected to office, he did serve the city as Fairmount Park Commissioner. It was during his service in this post that he enlarged Disston Park to cover five city blocks. Separate from this post, Hamilton Disston was the leading politician in Tacony. With Magistrate Tom South, he ran the town. South also sold real estate and was paid by Disston to handle the family's real-estate transactions in Tacony. Private Ledger No. 2 shows a salary of $3,784.57 paid to South for the year 1880. If this had been publicly known at the time, it would have been difficult for Hamilton to claim, as he did, that town justice was independent of Henry Disston & Sons. In any case, it is clear from these activities that Hamilton loved politics and excitement more than his job as chief manager of a large saw company.[2]

During a trip to Florida in 1877, Hamilton was moved by the vastness of the state and impressed with the opportunities it presented. Yet Florida seemed far away the next year, when his father died and he was thrust into the presidency of Disston & Sons.[3] With a group of investors, Hamilton bought more than 6,000 square miles of Florida land and went to work trying to make the land more desirable for settlement. He moved to Florida for a time, cleared land of water by building hundreds of miles of canals, and established the town of Kissimmee. He even had a Mississippi paddleboat brought to Florida, christened it the *Mary Disston* after his mother, and established the state's first packet line. He began to grow sugarcane and fruit trees along the canals, to create products for export. This would be the business from which a new community could be built, just as Tacony was built on the saw business. Hamilton was too much the adventurer to operate behind a desk in a factory.[4]

Unfortunately, unlike his father's similar venture in Atlantic City, the Florida investments were not a financial success. Hamilton's canals and steamboats were rendered impractical as the railroad became increasingly available to open up the interior of the state. Having invested heavily himself, Hamilton had enticed many powerful Philadelphians to do the same. As foreclosure on a $1 million loan approached, Hamilton took his own life to save his family's reputation.[5]

This left the saw works in the hands of Hamilton's brothers—William, Jacob S., and Horace C. Disston—and his half uncle, Samuel Disston. Samuel's energy and foresight were particularly important to the development of the firm during these years. He traveled extensively, making contacts and setting up business offices all over the nation. He was particularly effective in working with lumbermen and succeeded in making the Disston circular-saw blade the most popular in the country.

The period between 1878 and Hamilton's suicide in 1896 can best be described as one of benign town leadership. Divided power, as practiced under Disston's partnership model, lessened the force of paternalism, since there was no single powerful leader giving direction. Nevertheless, paternalism survived in the form of restrictive deeds and the management of the Mary Disston Estate.[6]

After Hamilton's death, William Disston took up the reins of control ably assisted by his brothers. William, born in 1859, started work at the factory in 1875 at age sixteen and served an eight-year apprenticeship in the shops. He became president of the firm in 1896 and held that post until 1915. Horace devoted himself to the steel department, and Jacob to the financial department. Samuel was appointed secretary and general manager.[7]

Their first consideration was to repay the $1 million loan made against company assets by Hamilton Disston in support of his Florida venture. William Disston secretly went to a James Stokesbury, a representative of J. P. Morgan, and borrowed the $1 million. Stokesbury granted the loan, and the reputation of the Disston family was preserved. This was important to the family, but a cash-flow problem would persist at the company for three years until the loan was repaid.[8]

During the years that Henry Disston was building his saw empire, he reinvested most of his profits in the firm. Now, the cash flow caused by the loan repayment and a growing list of Disston family members dependent on the profits from the saw works reduced the amount of money available for investment in town philanthropy. Hamilton Disston had incorporated the firm in 1886, and the stock was distributed among family members. Those who did not work at Disston (mostly the women of the family and the grandchildren) received equal shares. Those working at the plant received shares according to their responsibilities, tenure, and rank within the company. Hamilton received the most, with Jacob S., William, and Horace C. equally dividing the remainder of the management shares. Mary Disston, Henry's widow, had vast property holdings in Atlantic City, Tacony, Frankford, and Bridesburg, mortgages on hundreds of Tacony and Atlantic City properties, and nearly 20 percent of the company stock. Also part of this estate were three sawmills in Atlantic City that produced six-month profits of approximately $10,000, and reimbursements from the Tacony Water Works. In all, Mary Disston's income between 1880 and 1890 averaged nearly $112,000 per year from these investments and as much as $60,000 in dividends from the company. With these funds Mary Disston pursued Henry's paternalistic plan with more than the capital-poor firm could muster. She continued to maintain the homes in excellent condition and provided land for a school, a library, and a Presbyterian church. It was by her bequest that the Mary Disston Estate was established and financed until 1943.[9]

Splitting the Disston family wealth left the dividends of the factory as the sole means of support for many of Henry Disston's grandchildren. The men of the family could supplement their dividend income by working for the firm. In fact, it was the custom of the firm for each son to apprentice in the factory for five to seven years before assuming a leadership role. The only exception to this rule among Disston's sons was Jacob, who attended business school at the University of Pennsylvania. Upon reaching the age of twenty-one, the men of the family were admitted into full partnership in the firm and given a task commensurate with their abilities. As part of the company tradition, each was given a party in the department where he apprenticed, the most memorable being the party given for William in the Sample Room at Laurel Street in 1880. It started at 4 P.M. and lasted until 9 P.M., and the best food and drink available were served to the workers in that department to thank them for training "a Disston man in the ways of the Works." This practice continued as the Disston male grandchildren came of age.[10]

For the females in the Disston family, there was no work at the firm, only dividends. Therefore, it was important that sufficient money be available through dividends to provide a life-style befitting their place in society. Even for those who married into families like the Wanamakers, the Nalles, and the Gilpins, money was needed to provide for the leisure activities of Victorian women. This meant that dividends had to be substantial. A record of dividends paid to the family exists in Private Ledger No. 4 for 1898. Paid on March 4 of that year was a 10 percent dividend with a 5 percent extra dividend. The estate of Hamilton Disston received $108,180; William Disston $71,400; and Jacob Disston $71,400. The remaining twenty members of the Disston family shared $109,020. When one compares this total of $450,000 in dividends to the family with the

$5,976.36 paid to charity in the same period, Henry Disston & Sons' reputation for charity within the community is somewhat tarnished.[11]

The company's financial records for 1909 indicate that the Disston family remained the firm's greatest beneficiaries. William and Jacob S. Disston received $47,600, Elizabeth Disston $27,180, and the estate of Hamilton Disston $32,240. Hamilton's sons, Henry ($20,290), Frank ($18,300), and Albert ($16,960), were not far behind. William and Jacob were paid, respectively, $10,000 and $12,500 for managing the daily operation of the factory, giving some idea of the value of the dividends in 1909.[12]

The wealth from the factory transformed the lives of the family. Disston's sons began to expand their interest beyond the factory and real-estate investments of their father. In a 1900 list of financiers published in Philadelphia, the Disston family was prominent. Henry Disston, son of Hamilton, and William Disston were directors of the Tacony Trust Company; Jacob S. Disston was president of the Third National Bank and treasurer for the directors of the Philadelphia Manufacturers Fire Insurance Company Limited; and Samuel Disston was director of the Northern Trust Company. Excess cash from the factory and investments made them ideal candidates as financiers. The switch from factory income led many third-generation Disston men to take business training in colleges. Nevertheless, the family tradition of having sons work in the factory persisted until 1940, when William Leeds Disston trained with the workers. The Disstons never indulged in the social extravagances practiced by other wealthy families of the day. Even the Henry Disston vault in Laurel Hill Cemetery reflects the family's desire to venerate their father rather than mere ostentation.[13]

Disston's investments in the community—streets, water works, electric plants, schools, churches, playgrounds, libraries, and, later, trolley lines—had always been profitable, either bringing direct dividends or encouraging skilled workers to come to Tacony. The settlement of the estate after Henry's death put an extra burden on the company to produce larger profits each year. This left less and less money annually to be reinvested in the company for new equipment, and no money for the community. To work effectively paternalism required recyclable capital reinvested specifically for the benefit of workers. Because descendants of proprietary capitalists like Henry Disston tended to become numerous, increasing geometrically after just a few generations, the seeds were being sown for eventual cash-flow problems, especially in a bad year when the business cycle turned downward.[14]

The impact and implications of this changing financial picture were neither understood nor suspected by the Tacony population. Even the social changes in the community went largely unconsidered and unrecognized. A souvenir program was prepared for the May 30, 1906, celebration of Tacony Week, containing pictures of major local institutions and emphasizing recreation, health conditions, the pure water supply, a local newspaper, churches, the new free library, the modern post office, and two local hotels. The narrative proclaimed "a paradise for the working man of moderate means."[15]

Paternalism in the period from 1872 to 1890 produced a closely knit village of uniform values and a small-town way of life. One has only to read a single copy of the *Tacony New Era* to recognize the community structure. No letter carrier was assigned to Tacony; everyone met each morning at the post office in the hotel next to the Delaware River wharf to pick up the mail. There was no dentist in town (a concern to all), and only one doctor. Entertainment could be found in only one place: the Music Hall on Longshore Street. The churches—Episcopal, Methodist, Presbyterian, Baptist, Lutheran, German Catholic, and Irish Catholic—were all within walking distance. Most men and their sons worked at Disston, and everyone knew everyone else. The flavor of the town was so English that the Washington Tea House was a hub of the community. The community's interest in boats reflected its dependence on the Delaware River. The main thoroughfare, Longshore Street, ran directly down to the river, passing through the Disston saw works. After 1900, however, this unity of culture and small-town atmosphere would change rapidly.[16]

The first sign of change was an influx of new industries into the community. The Disston saw works would not remain the only industry in town. Disston had purchased 300 acres of land west of the Pennsylvania Railroad tracks between what is today McGee Street and Princeton Avenue extending west almost to Frankford Avenue. This farmland was inexpensive and easily purchased from farmers, who themselves realized a large profit from the deal. Deed restrictions prevented this land from being used as a factory site, but ample riverfront land suitable for industrial sites was still available and was bought up by other industrialists. Between 1881 and 1900, Taconyites witnessed the growth of six manufacturing companies in their neighborhood. Each added jobs and encouraged new people to settle in the town.[17]

The Tacony Iron Works opened in 1881 on land directly south of the Disston plant on State Road. Tacony by this date already had a number of skilled steelworkers from Sheffield. In its day, Tacony Iron Works was noted for its skilled workers. The structural iron work for the dome of City Hall and its mammoth statue of William Penn were made here during these decades. Tacony residents watched the progress of Alexander Milne Calder's statue with pride and delight. Its forty-seven pieces lay in the open field next to the plant, where many of the town's children played among them. These pieces were moved individually from the Tacony Iron Works to City Hall Square by Levi Eldridge's dray, a special sled for carrying heavy loads, and a team of horses driven by Taconyite Nick Tomlinson. He remained a local hero years after the event. Work on the Penn statue was not enough to guarantee financial success for the company, and in 1909 it went out of business.

The Tacony Iron Works building sat vacant until 1914, when the Lubin Film Studio at Eighth and Market Streets selected the site to stage a fire for a new film, *Gods of Fate.* The building was set on fire for a scene to be filmed with the stars escaping from a window. Unfortunately, the fire spread and the building burned to the ground.[18]

Another company that moved into Tacony to take advantage of the location and workforce was the Erben Search Company worsted mill. A factory was built south of the Disston plant, directly on the Delaware River, in 1885. Not a competitor for the existing labor force of the community, Erben Search sought unskilled textile workers, both women and men. In 1900 Erben changed partners and the plant became Erben-Harding. By 1910 it had gained a reputation within the community for hiring newly immigrated Italian and Polish women from nearby Bridesburg. Given the nature of the work and the limited number of employees (never more than 200, most of them women), the firm's impact on the private lives of the people of Tacony was minimal.[19]

Frank Shuman opened the American Wire Glass Company on land owned by Disston Saw in 1891. Such accommodations earned Disston revenue from the use of the ground and also fees for the use of the wharf. Moreover, the investors in Shuman's firm included some members of the Disston family. To everyone's satisfaction, the firm made money, and later a permanent plant was established two miles away at Comly Street and State Road.[20]

Frank Shuman built an inventors' compound at Ditman and Disston Streets in 1894. It was at this location that he experimented with the sun as an energy source for a low-pressure steam engine. After perfecting the engine in Tacony between 1907 and 1911, he sold his engine to the Egyptian government. Shuman's first functioning solar engine was used to irrigate land adjacent to the Nile. Until his death in 1918, Shuman remained Tacony's most notable private citizen.[21]

In 1894 Delaney & Company built a glue factory at Cottman and Milnor Streets, north of the Disston site. The smell from the glue-making process, and the working conditions, made employment in the plant less than desirable for most Taconyites. Nevertheless, the plant offered work to residents who could not find it elsewhere.[22]

The same year, the Luther Martin Lamp-Black Works opened a plant next to the Delaney glue factory. By 1900 the Martin Works was one of the oldest and largest lamp-black manufacturers in the United States. Lamp-black was used in making paints and printers' ink, an important product for Philadelphia's burgeoning printing and publishing companies. It is a fine soot of pure carbon formed by the imperfect combustion of highly carbonous substances and was produced in factories by using tar in specially constructed kilns. Visitors to Tacony often commented on the black-faced men returning to their homes from the factory each afternoon. However, like Delaney, Martin offered unskilled and undesirable work, compared with that available at Disston.[23]

In 1902 the Ross-Tacony Crucible Company moved to Tacony. Originally established in 1871 on Sixth Street below Jefferson, Ross tried to capitalize on Disston's need for crucible pots for making steel. The steel culture of Tacony made the new location ideal. The firm remained small and made Plumbago Crucibles for melting steel, brass, gold, silver, and other metals well into the twentieth century.[24]

In 1910 Gillinder & Sons Glass Manufacturers, known throughout the world for its high-quality glass, opened a plant on Levick Street below State Road. William T. Gillinder first established the business in 1861 in Philadelphia, and after his death the business was continued by his sons, James and Frank, until 1893, when the firm incorporated. The largest manufacturers of gas and electric glassware in the United States, Gillinder & Sons employed as many as 225 people at the Tacony plant.[25]

The France Packing Company, located along the Pennsylvania Railroad tracts below Unruh Street, produced various kinds of steam-packing for water, air, and gas acid boilers. Also important were the company's automatic compression grease caps and other general machine parts. These products were important for the new high-pressure steam engines. The company was opened sometime in the 1890s and advertised having a superior product called France Metallic fibrous packing. It located in the Tacony area because its products were crucial to the manufacturing plants in the area and a welcome addition to those that had steam engines.[26]

Each of these companies brought jobs to the residents of Tacony, increasing the number and kinds of work opportunities available. It made Tacony the place to go in 1900 if someone was seeking work. And go they did—by 1900 the town had grown to twenty times its 1870 size. Besides work at the factory, streets had to be laid, homes built, and water provided, all of which required labor beyond that provided by the resident English skilled workers and Irish laborers, most of whom already had jobs in the factories of the town. This opened the door for other ethnic groups, and so it was that Italians, Poles, Jews, and blacks came to Tacony for work. As the number of industries increased, the population of the town grew rapidly and ethnic neighborhoods developed.

These neighborhoods, besides being based on ethnicity, also reflected economic and social status in the community. The larger houses, mostly single homes, could be found on Disston Street not far from the train station. In the early days of the town, these homes were occupied by high-ranking department supervisors, foremen, and successful local bankers or builders. Those with the least wealth settled to the southwest, near Marsden and Unruh Streets. The skilled workers lived east of Torresdale Avenue around Knorr Street. The neighborhoods lowest in economic status were the Italian section in the northeast and the black neighborhood on Tacony Street. Status and wealth were also reflected by the houses themselves—whether they had plain or stained-glass windows, porches, or impressive or modest entrances.[27]

The ethnic change in the town was accelerated by a second trend: improvement in transportation. Electric trolleys were in vogue in the city, and the addition of lines to outlying districts like Tacony made them more accessible. Trolleys decreased cost and riding time for worker-commuters from Kensington, Frankford, Bridesburg, and Holmesburg to Tacony's factories. Tacony was no longer an isolated riverstop on the outskirts of Philadelphia, but was being drawn into the newly forming metropolis. The industrial workforce for Disston and the other companies could now be drawn from the city itself. This changed the physical structure and appearance of the town.

During the years 1872–1890 the community's basic transportation to the outside world consisted of horse and carriage, Delaware River steamboats, and the Pennsylvania Railroad, but none of these was useful on a daily basis for the average worker because the cost was prohibitive. Therefore, workers tended to live within two miles' walking distance of Disston's factory complex. Jane Marsden (daughter of master smelter Jonathan Marsden) was the first girl from Tacony to attend Girls' High School, then called Girls' Normal School, in Center City. To reach school, she traveled by boat down the Delaware, then boarded a horsecar. Earlier she had attended the public school built by Disston at Longshore Street and State Road, and later she taught at the Henry and Mary Disston Schools (as did her daughter, Ellen Marsden Rodgers). Philadelphia city life remained removed from this sleepy industrial community coexisting within its boundaries. But the electric trolley radically changed this relationship and expanded the area from which workers could commute on a daily basis.[28]

The most forceful politician to come from Tacony, and the person responsible for improved transportation to the area, was Peter E. Costello. Born June 27, 1853, in Boston, Costello came to Tacony in 1874. He lived in a boardinghouse and worked at Disston for eleven years. In 1888 he went into the construction business and built many of the neighborhood's homes. Some of his more elaborate constructions were Frank Shuman's houses on Disston Street between Marsden and Ditman Streets, and Costello's own home at 4700 Disston Street. The support of the Disston family greatly helped Costello's political career. As a developer, he became friendly with Thomas South and the Disston family. Homes and living space were still important to the Disston family, though in these days they favored home-building that did not cost the company money. Costello also organized the Suburban Electric Company in 1891.

Home-building and utilities led to Costello's interest in the development of the town. The first trolley line into the town—the Holmesburg, Tacony & Frankford Railroad Company—was founded in 1901 by Costello, who raised the funds for the line from local investors, including the Disston family. This privately owned line, with its small cars and Toonerville appearance, ran on one track from Holmesburg to Tacony to Frankford via Rhawn Street, State Road, and Orthodox Street. Dubbed the "Hop, Toad, and Frog" line, it remained locally controlled and financed throughout its existence, tying together the three largest communities in the far northeastern part of the city and giving the people of Tacony, for the first time, access to Center City by rapid transit. Nevertheless, the line's most important feature was that it provided transportation to the Disston plant and the Frankford Arsenal for these communities.[29]

Fortunately, a record remains of what it felt like to ride the "Hop, Toad, and Frog" line to work each day. A poem by Lou Gehr, telling of a ride to Disston in March 1918, compares this route to the Toonerville Trolley cartoon of that day. Line by line we are transported back to a time when trolleys offered a new daily experience for the working public:[30]

THE STATE ROAD LINE, ALIAS TOONERVILLE

We board a car at half past six
  On the corner of Bridge and Tacony
We're forced to stay there and stick,

Until it drives us loony.
And when we've laid there 'bout 10 minutes
The crew begins to stretch,
The motorman slowly dons his leather mitts,
(the heatless, pitiless wretch).

We've just about got going,
On our way up the line,
When behind us he sees a-running
For the company another dime.
The skipper stops the car in haste,
To let two more get on her,
No part of space does go to waste,
As long as there's four corners.

We're packed together like sardines,
There's not much room to breathe;
But they keep packing us right in
Until you're like a sieve.
Finally the boat begins,
Proceeding to the switch,
Where the conductor starts right in,
To gather in the chits.

We do not wait very long,
'Bout five minutes or ten;
Then on our way we soon are going
Up State Road again.
We're not going but a short time
When to a stop we come,
The conductor has taken in a dime,
Which he claims is bum.

An argument right there does flow,
O'er that mutilated coin,
And after ten more minutes go,
Again the boat starts going.
By the time we get to Comly Street,
Our feet are almost froze,
Our faces are as raw as beets,
Our noses just like red roses.

The usual fifteen minutes' wait
Here always starts a brawl;

The car to come is always late,
  And sometimes not at all.
Again we've started on our way,
  We pass thru Wissinoming,
A "Hell of a line," you'll hear 'em say
  On almost every morning.

We reach the car barn where a boat
  Was wrecked from trouble of motor;
They jam on top of us their load,
  And then we continue further.
Up Levick Street hill we climb,
  Straining every bolt and chain,
Should power be shut off at this time,
  We'd go crashing back again.

Over the tracks at France Mett. Packing,
  We go smashing with a jar;
A shock that makes your spine go crashing;
  And shatters windows in the car.
Then on the last lap we do go,
  At Knorr Street we do lurk,
We disembark nearly fagged out,
  A half hour late for work.

On August 26, 1903, a second trolley line was completed. This one traveled along Torresdale Avenue and connected in Frankford with the "dummy" steam trolley that went to Center City. This new route was run by the Rapid Transit Corporation of Philadelphia. The trolleys were larger, faster, and more modern, and they ran on time. There were two tracks, which meant no waiting on a side track for the car coming the other way, and the quality of the tracks made the ride smooth and comfortable. The No. 58 trolley (later the No. 56) on Torresdale Avenue eventually ended the community's dependence on steamboats and the railroad for transportation to the city center. It also shifted Tacony's axis away from the Delaware River.[31]

Tacony's major thoroughfare had been Longshore Street, which was near the main gate of the Disston saw works, contained the main waterline for the town, and became the commercial and social center of the community. The bank building, the Music Hall with its meeting rooms and library, the savings and loan, the department store, the movie house, the police station, the firehouse, the park, and the commercial district all stood along the seven blocks between State Road and Torresdale Avenue.

Three years after completion of the trolley line, the first shopkeeper (grocer Edward Darreff) moved his store from Tulip Street and Longshore to Torresdale Avenue. Darreff's son, Edward Jr., remembers how everyone on Longshore Street cautioned his father not to move out "into the woods." John Rapp, president of the bank at the time, told Darreff that although he had the credit to get money for his new store, "You're a damned fool for moving out there." Within five years Darreff had twenty more stores around him as other Longshore Street businesses followed suit.[32]

Located approximately eight blocks west of Disston's plant and the river, Torresdale Avenue had by 1930 supplanted Longshore as the district's main street. The movie house (the "New Liberty"), followed by shops and the bank, moved to Torresdale Avenue near Longshore Street to form a new axis for the community. This event both physically and symbolically turned people away from the river. The focal point of the neighborhood was no longer the Disston saw works. Facilitated by rapid transit, the powerful draw of Center City had within ten short years made the old economic provincialism obsolete and built a new hub for the community. Transportation had changed worker dependence on the Disston company and restructured the ethnic population of Tacony.[33]

City life, especially in Philadelphia, had long comprised many different ethnic groups, which took the trolleys to Tacony seeking work. Italian immigrants settled at the end of the line at Cottman and Torresdale Avenues after helping to lay the tracks. Most of these original Italians were from South Philadelphia's Ninth and South Street area, attracted by the openness of the northeast. A group headed by Giorlando Tumolilo, Vincent Troilo, Vincent Tortorello, and Euclide Jasca had formed the United Independent Italian American Club of the City of Philadelphia on February 5, 1896, in South Philadelphia, with the avowed purpose of educating and improving the lives of Italians in America. In pursuit of that goal, they moved family by family from crowded rowhouse districts to the open farmland at the end of the No. 58 trolley line at Torresdale and Cottman Avenues. By 1912 the same group had purchased an old German Catholic farmhouse and opened a club in Tacony. With bocce courts in the rear, a picnic area on the side, and a main hall for large parties, it became the center for their community's social life. Other rooms served as temporary quarters for new arrivals to Tacony from South Philadelphia or Italy. Men found ample work in Tacony building homes for the fast-growing community, but such Tacony institutions as the Music Hall, the Tacony Club, the Disston factory, and even St. Leo's Catholic Church were hostile to their presence. Despite their nearness to

St. Leo's—a predominantly Irish church—in 1917 they opened their own church school, Our Lady of Consolation, on Princeton Avenue.[34]

The Italians settled north of Princeton Avenue and away from Disston's deed-restricted land. They openly served wine at their club and had weekly festivals in the yard outside it. Most of the early Italians worked in the least desirable jobs, at Delaney's glue factory or Martin's lampblack factory. Older Taconyites still tell stories about Italian men coming home from Martin's factory with dirty and blackened faces. Eventually, a few Italians gained work at Henry Disston & Sons, but they were at best only mildly influenced by the company's paternalism. The Italian community fostered its own social and ethnic organizations, which overshadowed those sponsored by the saw works.

A severe labor shortage at Henry Disston & Sons in 1918, created by men leaving to serve in World War I, brought yet another group to Tacony. The Pennsylvania Railroad, to meet its wartime demand for labor, carried many blacks north free of charge. William D. Disston did the same thing. Sending Harry and Edmund Whittaker to Remington, Virginia, he was able to entice 400 black workers to come to his plant. Some were given housing on the 6900 block of Wissinoming Avenue, owned by Mary Disston and dubbed the "Mary Seven," and others lived out of jerry-rigged boxcars set off a siding and owned by Disston. The arrangement worked well for Disston but not so well for the black workers, who lost their jobs to the returning servicemen when the war ended. Nor did the minstrel shows sponsored by Disston & Sons help to ease the black population into the mainstream of neighborhood life. Nevertheless, some blacks continued as laborers at Disston and built the Star Hope Baptist Church in the community. Like the Italians, blacks were marginal to Tacony's social life. Isolated and economically disadvantaged, both groups suffered the consequences of being on the wrong side of the tracks in this still strongly English town.[35]

In the 1920s, Jewish immigrants began moving into the community in small numbers—at most twenty families. They worked for others or opened stores on Longshore Street throughout the decade. Leaders in the group were the Rubin brothers, proprietors of a department store. Hyman Rubin, son of one of the brothers, studied law and returned to the community to set up practice. But he was the exception. Most Tacony Jews worked all their lives in the neighborhood's stores. Their business included Joe Rosenwald's grocery, Myra Zier's shoe store, David Brees's lady's shop, and Ike Guggenheim's dress shop. Only a few Jews worked at Disston. The longtime residents of Tacony considered the Jews clannish, but as Hyman Rubin's daughter Judith explained, "The deepest disappointment to my father, who loved Tacony, was that he was denied entrance into the homes of his Tacony friends." Prejudice was subdued in Tacony, but it was there.[36]

The exact population of each ethnic group is difficult to determine, but the census of 1880, a local publication; church populations; and a 1920 study provide sufficient information to speculate on population sizes in 1880, 1906, and 1920 and to give some idea of how the community changed over a forty-year period (see Table 5).

An incident during this time tells a great deal about racial attitudes and values in the community. The Aces football team, made up of the community's Protestant boys, stored equipment in a garage next to George Kett's ice house on Unruh Street below Hegerman that also served as their clubhouse. The team's big game on Thanksgiving Day against the Catholic team at St. Leo's was a month away, and the Aces were bored. A group of them met at the garage on Mischief Night in 1921 and were discussing what they could do for excitement that night. On the spur of the moment, they decided to scare the blacks from Remington, some of whom had stayed on after the war in the Wissinoming Street section of the town, working at Disston or doing odd jobs.

Lumber from the yard was nailed together to form a cross. The boys took the curtains made by Ted Shoutline's girlfriend from the window and wrapped them around the cross. Pat Carson was sent to buy a half-gallon of gasoline at the local station while the club members, about thirty in all, drank soda and ate candy. There was a small fire in the center of the garage, and Buck Omrod took a cup with gasoline and threw some drops of gasoline into the flames. The flames flared up and the boys laughed. On the next attempt, Buck's hand slipped and the full cup of gasoline went into the fire. An explosion followed. The garage door flew open, flames filled the building, and all thirty boys got out in a few seconds. They ran away, but not before one of them called the fire

Table 5. Ethnic Population in Tacony, 1880–1920

	Other	Italians	Irish	English	Germans	Blacks	Jews
1880	2%	0%	25%	55%	18%	0%	0%
1906	2	5	29	42	21	0	0
1920	5	10	27	39	17	1	1

SOURCES: Interview with Marguerite Dorsey Farley, June 14, 1990; U.S. Census, 1880, Philadelphia, 35th Ward, Philadelphia Free Library, Logan Square; *Tacony Souvenir Program* (Philadelphia, 1906); "A List of Churches and Populations in Tacony, 1920," in *Tacony* (Philadelphia: New Era Press, 1920) (the latter is in the Tacony Public Library).

NOTE: The percentage figures quoted represent only a best estimate from the available documentation and interviews.

company. Fire companies came from Tacony, Holmesburg, and Frankford, but there was no water in the fireplug on Unruh Street. The garage burned to the ground. This prompted an investigation by city officials of the poor water pressure in the community, which led to the sale of the Tacony Water Company to the city. Tacony began, for the first time, to use water from the city's new Torresdale plant. Officials were unable, or uninclined, to investigate what caused the fire.

The following night a meeting was called in Bob Bachman's house at 4506 Longshore Street. Herb Pfeffer, an older sponsor of the club, sat them all down and reprimanded Buck Omrod for starting the fire. Then he told the scared group of young men to keep quiet about what happened because the club could be in trouble if word got out. Nothing was said to the group about their objective—namely, to burn a cross in front of the Mary Seven. This story illustrates the submerged, internalized racism of the community and also explains why this behavior caused few riots. Minority populations were too small and thus had little influence on the accepted values of the majority population, a population that feared disorder but not being labeled racist.[37]

Nevertheless, improved transportation and pluralism brought unsavory types from the city. Vendors lined the streets outside the plant on payday. A poem by one Disston worker describes the scene at the Knorr Street Disston gate at the end of a workday in 1918.[38]

THE DISSTON MIDWAY

Every week-end, at Knorr Street Gate,
The fakirs, they do congregate,
Give to us a great Midway
Helping us to spend our pay.

Hear them holler, hear them shout
In telling you their wares about
They holler so, they almost choke,
Getting you to buy their dope.

They have a cure for every ill,
Sell to you a patent pill,
If it's corns that you may have
They recommend their Wonder Salve.

King of them all is Mr. Kirk,
From your back he'd take a shirt.

Sells to you so very cheap,
Beyond his rep, for him to cheat.

There's Rough and Ready, a Western man,
Hails from Texas, by the Rio Grande.
Sells good oil, from a snake,
From your body pains will take.

"Hot peanuts! Hot Peanuts!" hollers Jack;
Sells each week about a sack.
The kids at home for them wait
And loudly cry if dad is late.

Watches, too, they have on sale,
Easy they do take your kale.
Always something home to take,
The fakirs have at Knorr Street Gate.

Taconyites were now less likely to know their neighbors. The town had grown large enough to promote anonymity. Italian, black, and Jewish immigrants, in small numbers, were settling in separate neighborhoods. Fear of a breakdown in community values promoted interest in social organizations to bolster family life and ensure ethnic stability and continuity. Deed restrictions were losing effectiveness as a means of social control in this growing and populous pluralistic society. The Italians were making homemade wine; the Tacony Club was serving liquor privately to businessmen; and stables for private businesses could be found throughout the community. Two legal public bars (Merz's and Holbart's), located outside the Disston land tract and next to the factory, openly served liquor to the workers. In the minds of the men of the town, new social control mechanisms were necessary if order and community spirit were to survive. A new Fathers' Association appeared to deal with these fears and tried to rekindle and restore the values of the "old town."[39]

The Tacony Fathers' Association was founded in 1917 and continued into the 1920s. Composed of about 400 of the town's leading citizens, it had the objective of "bettering living conditions" in the town. In a growing town that now housed more strangers than friends, the association declared that Tacony was due "more light, more police protection, better drainage system, better school-house facilities, through trolley cars to center city without charge, and numerous other equally important things."[40]

Its first meeting was held in the Tacony Library in December 1917. Organizer John H. Glenn urged that "every man who is interested in the welfare of our boys be there." Speakers had been invited, and the "Temple Quartet," led by well-known Disston foreman Elwood Gebhardt, would perform. Glenn made it clear that admission was free and appealed to the pride of the residents: "Other nearby towns boast of their Fathers' association—why can't we?"[41]

Meetings were held on the third Wednesday of the month at the bank building controlled by Disston & Sons on Longshore Street. Advertisements for meetings emphasized programs with interesting speakers and entertainment, with a special appeal to self-interest: "It will pay you to join and help make this movement a bounding success."[42]

At the height of the war, in 1918, war heroes became the association's main attraction. Lieutenant Heintzman of the Canadian Infantry, one of the first 33,000 troops sent to Europe, spoke for forty-five minutes, giving graphic descriptions of the battlefields of France. Later, Sergeant MacNiff, one of the eleven men who had returned from Pershing's army, told what American soldiers were doing in France. Appeals were made for the audience to buy war bonds, and McIlhan's orchestra provided music at the end of the program. After the war, the Fathers' Association actively supported the founding of a Tacony American Legion Post. The appeal was simple. "Help the boys who saved our homes." Named after William Oxley, the first Tacony boy killed in France, the Legion would eventually supersede the Fathers' Association.[43]

A second community enterprise focused on controlling the town's "spirited boys" through sports. The history of sports in Tacony, which goes back to the early days of its industrial development, points up both continuities and changes in the relationship between the paternalistic company and the community.

Although a variety of teams would eventually play under the Disston company banner, the original teams were organized by townspeople. The mainstays during these early years were baseball and soccer. Henry Disston brought some local cricket players to Tacony in 1876–1877 and laid out the community's first playing fields between Longshore and Disston Streets on the west side of the railroad tracts. He played in the games himself when he was fifty-six years of age, against local players from Tacony and Holmesburg. Shop teams developed in the 1880s, and eventually a baseball league was begun between the file, saw, and steel shops. After work, games were played all summer in Disston Park. Not highly structured, these leagues allowed for the development of local talent. In 1889 Frank B. Fisher, a local sportsman and member of the Tacony Club, organized the first Tacony team. A year later Fisher's team officially became the town's team and was renamed the Tacony Athletic Association. Contributions of $5 each from sixty members of the community, mostly members of the Tacony Club, supported the team. The association rented a plot of land at Unruh Street and State Road at a reduced rate from the Disston family, filled and leveled it, and constructed a high board fence with grandstands. The first year brought enough money from paying crowds to meet all expenses and give each player a salary.[44]

In succeeding years, crowds turned out by the thousands to see the games. Tavern owner Charlie Merz and Disston superintendent Albert Butterworth Jr. now offered full support to Fisher and his team. Better players were signed, and Tacony became known as an avid baseball community. Billy Seed, Mace Ploucher, and Jack Glenn became household names in Tacony. During the week, seven of the nine men worked at Disston, and Assistant Coach Butterworth arranged for them to get off early from work two nights a week for practice.[45]

Henry Disston's death in 1878 did not end the family's interest in the Tacony teams. The new president, Hamilton Disston, was known to come to Tacony on Saturdays just to watch the games. The next week in the shops he would personally congratulate the men on their play. By the 1890s, Tacony fielded one of the finest semi-professional teams in Philadelphia, thanks in great part to the support of the Disston saw works, which provided jobs and time off for practice. Games were played in the coal regions of Pennsylvania and in Atlantic City. It was not unusual for the team to bring hundreds of Taconyites with them on these weekend train excursions. In 1897, 1898, and 1899, with pitcher Billy Seed leading the way, the Tacony Athletic Association team became amateur baseball champions of Philadelphia. This achievement led a number of players to go into the professional minor leagues. By 1901 there were not enough players available for the Tacony A. A. to field a team.[46]

The life of Billy Seed illustrates some of the subtleties of small-town paternalism. The son of a immigrant English engineer, Billy was born in England and came with his family to Tacony in 1876. He began working with his father at Disston when he was 14 years old. He learned the game of baseball on the sandlots of Tacony, becoming the pitching star of the Tacony team between 1889 and 1900 and often pitching right-handed against left-handed brother Dick. Married in his twenties, Billy continued to travel with the team on weekends throughout the summer. Games were played on Saturday or Sunday. His normal routine during the summer was to work from eight to five o'clock at Disston, practice two nights a week until 8:30 P.M., and go with the boys to Charlie Merz's saloon afterward. It was here that Billy learned to drink. He named his second son Merz Seed (the first was Bill Jr.). As we have seen, sandlot players like Billy Seed received special treatment at Disston. Albert Butterworth rearranged schedules to fit games and saw that an occasional late-night drunk and the subsequent hangover did not get a man fired.[47]

Despite these privileges, Billy Seed left the Tacony team to play for the Boston Red Sox in the New England League in the 1890s. He spent three summers away from his family, only to return each winter to the job that awaited him at Disston. His nickname in Tacony was now "Sox," and his fame as a professional athlete made him the idol of the boys of the town.[48]

Billy drank too much as he aged. It was common for him to be sent home from work once a week because of drinking. Yet Henry Disston & Sons never fired Billy. Loyalty to Tacony's greatest baseball player was something everyone at Disston understood, including management. Disston's paternalism was not aimed at profitability alone, but had a large cultural context: Firing Billy Seed would have caused shock and trouble within the community, so he was kept on for many years as a "guest" worker.[49]

Every Christmas, Bill could expect a card from his good friend Henry Disston, grandson of the founder. Finally fired by a supervisor after forty-eight years of service, Bill appealed to the Disston family and was given a job as a laborer in 1944. This demotion disappointed Bill, who began to drink more. Eventually he retired and was awarded a pension of $15 a month because of his low job status.[50]

During the years between 1903 and 1917, soccer became the sporting mainstay of this community, which was still made up largely of Englishmen. In 1901 restaurant owner Harry Lister and recently arrived Englishman Sam Needham formed the community's first soccer team. Needham, a boarder in the household of master smelter Jonathan Marsden, lined up the players and laid out a field on the lot in front of St. Leo's Church at Unruh and Keystone Streets. Lister raised the money for referees and uniforms. No Philadelphia league existed at the time, so Needham arranged games with the Oxford team of Frankford and a team from Trenton, New Jersey. In contrast to baseball players, soccer players participated without pay and paid their own expenses to away games.[51]

From 1907 to 1914 Tacony had the finest soccer team in Philadelphia. Needham, coach and architect of the team, recruited players from Canada and England. It was not unusual for him to show up at the Disston personnel office with a newly arrived English worker who happened to be an outstanding soccer player. Hector McDonald, the most notable and skillful player in the city, was among the recruits. In 1910 the Disston team was Eastern Champion and went to St. Louis during the Christmas holiday to play for the American Cup (they lost, 3–2). The following year

Disston won the Pennsylvania championship but was defeated by Paterson, New Jersey, on the third replay—the first two games being draws.[52]

Before schools and colleges sponsored teams, community and factory teams were the backbone of amateur sports organizations. It was not unusual for local factory teams to play the colleges of the day. In May 1919 the University of Pennsylvania played the Disston Athletic Association soccer team to a 1–1 tie. According to contemporary descriptions, the game was played hard by two equally talented teams.[53]

Despite the Fathers' Association and private athletic teams, however, paternalism as originally designed was in trouble in Tacony. The nuclear model had exploded. New factories gave residents a choice of work; trolley lines brought in workers from other parts of the city; financial considerations and disbursements to Disston family members precluded spending large sums on the town's needs; separate ethnic communities spawned different ways of life; and private efforts by the community itself were to prove inadequate. After 1900, paternalism as a method of controlling the workforce was being undermined by forces outside the community. The strongest remaining vestige of Henry Disston's paternalism was the system of land-deed restrictions and family ownership of worker homes as prescribed in the will of Mary Disston.

In the factory, significant events took place as the community and the Disston company moved away from uniformity and paternalism. First, the period between 1902 and 1916 marked the largest building program in the history of the company. During these four years, the Disston company spent more than $1.5 million on equipment and buildings. In 1902 a mill was built to cold-roll handsaws, a process superior to the hot-roll process formerly used. In 1907 a wood mill was installed so the company could purchase applewood logs rather than finished planks. In 1909 a machine shop was built. In 1910 a new machine knife and jobbing building was constructed, as well as a pattern shop, a blacksmith shop, a new cold-rolling mill, and a concrete warehouse. In 1911 a file shop, a time office, a garage, and a low-pressure steam turbine generative plant were introduced. Business was increasing, and the Disston management was positioning itself for the increased business that would come with the outbreak of World War I. In 1916 a new and larger file shop was built for forging files, and four dry kilns for drying saw handles, as well as a fully equipped dispensary with a resident physician in charge.[54]

A second event was the training at the factory of the leadership that would guide the company for the next half-century. S. Horace Disston began his three-year apprenticeship in September 1899; William D. Disston began his on January 6, 1908; and Hamilton, son of Jacob, started in September 1908. Each would go on to be a CEO of the company during the twentieth century. Their decisions and judgments gave direction to the firm and kept it the leading saw company in the world. They studied the processes and operation of the saw- and steel-making departments, which performed the actual work of shaping the crude material to the finished product. It was the way Henry Disston had trained his sons, and it was the way sons were expected to learn the business.[55]

CHAPTER

5

Paternalism to Industrial Welfare, 1909–1929

BEFORE WORLD WAR I, the idea of industrial welfare attracted the support of social reformers, civic organizations, and universities. The basic goals were to improve employee-owner relationships, socialize immigrant workers to industrial mores, defuse employee discontent by providing amenities, and foster workers' loyalty to one another and to the employer. The University of Chicago, Yale University, and other institutions introduced courses in industrial welfare. Industrial welfare won the eventual endorsement of President Theodore Roosevelt and later President William Howard Taft. President Woodrow Wilson supported the concept as a means of stimulating production during the war. So too did Henry Disston & Sons.[1]

"Industrial welfare" was a term that became identified with factory owners who sponsored community-centered activities that encouraged community and worker loyalty to the company. Sports teams, lecture programs in the community, and company newspapers filled these functions. Today remnants of industrial welfare programs are written into union agreements that protect the rights of individual workers. Such contracts have replaced the voluntary programs begun under company industrial welfare policies. The difference is that now the programs are guaranteed, and pensions, seniority rules, working conditions, stock options, medical benefits, and other fringes are not left to the whim of the employer. Company newspapers, teams, social activities, picnics, and trips to amusement parks have for the most part disappeared. As for other benefits formerly provided by companies, most workers seek these services under government plans.

Once, however, a company's best means of controlling its workers and ensuring their stability and loyalty rested in industrial welfare programs that could be attributed directly to the beneficence of the employer. The Tacony/Disston experience is a microcosm of the movement from paternalism to industrial welfare to the welfare state that took place throughout the United States between 1880 and 1960.[2]

The industrial management adaptations necessitated by the shift from paternalism to industrial welfare were almost imperceptible in Tacony. This is no surprise, since the ideas of paternalism were not far removed from the ideas of industrial welfare. Nevertheless, there was a significant shift from owner direction to worker direction. During the years paternalism was in vogue, the Disston family took great interest in the development of the town. Building homes, and providing schools, churches, a gas company, and a water supply, the company acted like a parent guiding children toward the good life. Much of what was done for the town rested in the Disston family's intuitive sense of what was right.

Industrial welfare promised a different and better way to manage the lives of employees. Intuitive feelings about what made a better worker were transformed into more-scientific and measurable programs. The production goals became more specific at the factory, and the individual nuclear family took on greater significance as a means of stabilizing workers' lives. At the factory, an employment office placed workers in the right jobs, a dispensary provided for their health needs, recreational opportunities were made available to all, training for specific skills was instituted, social events in the evening became part of factory life, a lunchroom was provided, and the company began to set money aside for a retirement fund. Although these initiatives by Disston signaled a movement away from paternalism, the workers and townspeople continued to have warm and respectful opinions of the Disston family and the company. Disston's focus on health and safety, recreation and sports, and social events organized by and for the worker, and above all its company-run newspaper for workers, allayed the uneasiness in the community caused by the social changes inherent in the town's rapid growth and increased pluralism.

Henry Disston & Sons had long produced the *Disston Crucible,* a monthly magazine that gave tips on the use of saws in general and advertised Disston Saws in particular. It played an important role in enabling Disston to corner the lumber market in the early part of the twentieth century. The *Disston "Bits": A Monthly Magazine of "Inside Stuff"* was first issued in July 1917 with the objective of stimulating and crystallizing "among the employees" a mutual spirit "of good will and co-operation." The drawings, editorial matter, and topics "will be entirely the work of Disston employees," management promised. "It will be more than a spokesman, it will be a clearing house for the ideas of individuals in the Disston plant." While the paper never attained this ideal, it did address questions related to worker safety, health, recreation, and sports, and, more important, served as a forum for propagating the ideas of worker welfare as perceived by management.[3]

A quotation from William Howard Taft in the second issue of the "*Bits*" establishes the ground rules for the unity management desired. If a man "does his work for the love of it, and not out of consideration alone for the result, he will serve his owners' interest best, for he will do his work well and thereby make himself indispensable to his employer; and when the time comes to choose a man for a higher position the choice will likely fall upon him who has done his work well." It made no difference that this was not a workable formula at Disston, where the family reserved the right to the top management positions. What was important was that the workers were supposed to believe it.[4]

The safety plan instituted in 1911 had not done all management hoped. The plan, like Taft's quotation, went beyond the normal concern for safety and attempted to link safety to attitude toward work. The workers were reminded that their very thoughts affected their health. Sickly thoughts manifested themselves in a sickly body; strong, pure, and happy thoughts built up the body's vigor and grace. To think well of all, to be cheerful with all, to learn patiently to find the good in all—such unselfish ideals were advocated by management. In spite of its simplistic approach, the plan did recognize the importance of a mentally healthy worker even during this relatively primitive industrial time. Management did not, in fact, achieve a rejuvenated workforce at the plant, but it did provide health care for workers injured on the job.[5]

The first dispensary, fully equipped and with a physician in charge was opened in 1916. The doctor who operated the clinic for more than thirty years was Augustus Valentine, who became known throughout the community as Disston's doctor. A blunt man who swore easily, he was popular with the plant workers. He was instrumental in establishing the company's campaign against infected wounds, first advocating treatment for every wound that occurred at the factory and later leading a successful campaign to eliminate the use of river water for handwashing at the factory because it was a source of infections. Dr. Valentine's yellow Studebaker was his trademark, and it was often seen streaking through the plant gate at breakneck speed, escorted by the police, to attend a plant emergency.[6]

The dispensary was put to good use during World War I. During the first year of America's participation in the war, the government disclosed that more soldiers were incapacitated by venereal disease than by all other maladies. Despite a reluctance in communities like Tacony to discuss such topics openly, an article in the company and community papers spoke candidly about venereal disease: "Do you know that prostitution is the cause of nearly all venereal infection? Gonorrhea and syphilis are 'camp followers' where prostitution and alcohol are permitted," warned the *"Bits."* Moreover, the greatest proliferating agent of venereal disease in the army was the influx of infected civilian recruits. To safeguard the soldier and soldier-to-be, all communities needed to be freed from sources of venereal disease. The medical department at Henry Disston & Sons had a doctor ready for consultation about any illness, including sexually transmitted ones.[7]

Treatment of industrial injuries, however, remained the principal function of the dispensary. The company used the newspaper to enumerate the causes of the most common injuries treated there. They were, in order of occurrence, cuts caused by handling materials, eye injuries, burns from hot metals, injuries caused by machinery, and infection caused by neglected cuts. Prevention was the key to avoiding accidents, and those with even the most insignificant injury were told to report to the dispensary. Foremen were to see that dressing of wounds took place at the appointed time. Individuals who neglected to have wounds dressed received letters at their home.[8]

Posters and signs were placed around the factory to encourage safety. The foremen were charged with getting jobs done safely, talking about safety procedures with the workers, keeping their areas orderly, stopping dangerous practices, and taking an interest in workers' injuries.[9]

Stories of accidents avoided by safe practices proliferated in the *"Bits."* In January 1918 Nick Rocco of the steel works was trimming some burrs from a piece of steel when a large piece struck his goggles. The glass was broken, but there was no damage to the eye. The editor of the *"Bits"* claimed that the goggles saved Rocco's eyesight, appealing to workers to "Ask Nick—he knows." A year later, workers were still being warned to wear goggles. If they were pitted, broken, or dirty, foremen were encouraged to send them to the safety department for repair, cleaning, or replacement.[10]

Slogans and poems reinforced management's view of safety. Common aphorisms were "A wooden leg is often the product of a wooden head" and "It is cheaper to keep well than to get well." A typical poem read:

THE COFFER

You don't believe in Safety First?
  Please give your reason why.
Do you believe in accidents?
  D'you want to see men die?
Pray, who gave you the awful right
  To prolong life or cut off life?
Or even harm a single hair
  Of a comrade in his daily strife
To get along and make a home.
  And live for himself alone.[11]

Even community entertainment was planned around the theme of safety. Dr. Valentine arranged lectures for the community, held at the Tacony Public Library and paid for by the company. One speaker was John S. Spicer, chemical engineer for the state Department of Labor, who complimented the company on its fine safety record. S. Horace Disston, then a vice president, followed, congratulating the employees on their support for the company's safety plan. To show that it was working, Disston presented data that indicated a decrease in worktime lost to injury in 1917. Two motion pictures followed: "Carelessness and Caution," which contrasted good and poor fire-fighting practices, and "The House That Jack Built," which showed the horrible results of careless habits. Six months later a lecture entitled "The Danger of the Fly to the Community" attracted an even larger crowd.[12]

The "*Bits*" continued to carry story after story about safety. Canvas shoes were not to be worn to work because they conducted excessive heat to the feet, which made them "tired," and because of the danger of stepping on nails or getting too near a hot fire. "The study of safety is the study of the *Right* way to do things," the paper announced.[13]

The company often stretched the concept of safety to encompass ideas that promoted production, as illustrated in the story entitled "Be a Booster for Safety." The article implored the workers not to keep their eyes on the foreman but to pay attention to the job, not to throw goggles or other items haphazardly around the shop, and not to litter floors. They were to mind their own business, stay at their assignment, do their best, consult the foreman, and refrain from knocking the boss. While these practices might tangentially have improved safety, they surely did far more to enhance production and a good plant atmosphere for management.[14]

In 1916 the company began keeping records of those injured while at work. Despite all precautionary rhetoric, the number of accidents increased from 2,539 in 1916 to 3,506 in 1917. A year later, the cases treated in the company dispensary during the same period increased from 4,824 to 10,726. The emphasis on vigilant treatment of infections, which increased the number of cases reported and the activity in the dispensary, likewise resulted in a decrease in the days lost from work because of accidents from 5,471 to 2,891. By May 1919 that number had dropped to 1,599 days lost, despite 13,739 cases treated in the dispensary that year. The success of this plan rested on the reduction of infection, a constant threat in all plants where the workers worked with steel.[15]

Another concern of management was that the new state compensation law not be "misunderstood" by the workers. Workers were cautioned that while the law provided for surgical coverage and financial aid for those injured at work, it offered no remuneration for pain, suffering, or permanent physical damage resulting from an injury. Disston's aim was "that cripples and helpless wrecks who were once strong men shall no longer be the by-products of industry."[16]

Clearly, promoting safety and limiting production costs were mutually compatible. The Disston plant record for safety and for efficacious treatment of injuries ranked among the best in the city and contributed to the workers' sense that the company cared about them.[17]

Sports were a second means of encouraging company loyalty and spirit. In a small community isolated from the big city, weekend games were valued entertainment. Although a variety of teams played under the company banner, the mainstays during these early years were baseball and soccer.

The Athletic Association was formed at Disston on December 14, 1917. Meetings of the company-supported Henry Disston & Sons Athletic and Recreation Association were held the last Friday of the month, first in the factory and later in the bank building, with Henry Disston (son of Hamilton and president of the firm) as chairman. In order to attract members to the meeting, a dance was scheduled at 8:00 P.M. after each monthly meeting. Caleb Purdy was named entertainment chairman, and noted community singers and the Disston orchestra, an organization consisting of Disston workers, were asked to perform. Committees were organized for baseball, soccer, bowling, basketball, volleyball, quoits, gunnery, groundskeeping, entertainment, and track. Girls' teams were offered in baseball and recreational dodgeball.[18]

The membership contribution was set at $1 a year for men and 50 cents for women, which helped defray the expenses of the organization. Membership was confined to Disston employees and entitled members to admission to all entertainment sponsored by the association, including dances and minstrel shows. Besides playing other industrial teams in the Northeast Industrial League, the Athletic Association organized competitions within the factory. In 1919 the steelworkers, file workers, office workers, and saw workers competed on the baseball field. Men's and women's track-and-field meets were introduced during the 1920s. During lunch periods, the company-owned recreational park adjacent to the factory was filled with women workers playing dodgeball and men throwing or kicking balls. Final scores and photographs from these events were published in the company newspaper.[19]

The Disston Minstrels were another vehicle management used to foster company spirit among the workers. Sponsored by the Athletic Association, these entertainments consisted of fifteen to twenty black-faced workers rendering songs and malapropisms for audiences packed into the company cafeteria. The sort of racism represented by minstrel shows was not an issue for Disston's white workers; its few black employees did not complain, perhaps fearing that they would be classified as troublemakers and forced to leave their jobs. Other Athletic Association programs featured dancing, singing, instrumental solos, and the Disston orchestra.[20]

The Northeast Industrial League was formed in 1920, pitting Disston against the nearby Barrett Chemical Company, R. H. Foerderer Leather Manufacturer, the Frankford Arsenal, Swartz Wheel, David Lytton & Sons, Miller Lock Company, and Fayette R. Plumb Company. Home games for fall and spring sports were played at the Disston company recreation field, while winter sports like basketball were played indoors at the Odd Fellows Hall in Kensington. A dance followed each game. Track and Field Day in August brought sports stars from all over the city to compete against Tacony's fastest man—the head of the new personnel department, Harry Dorsey.[21]

Fielding a winning team was of major concern to the workers and management. Good morale and company loyalty were a by-product of victory. It was not unusual for management to recruit someone from another company because he was a good baseball or soccer player. Englishmen with soccer talent sought employment at Disston each year, entrenching Disston as the best team in its league from 1900 to 1925. Even the hiring of a coach became important company business. In May 1918 the announcement that Harry Frederick had been appointed the new manager of the baseball team brought favorable comment from all. Harry "knows the game from A to Z and is a leader when it comes to the business of handling men. With proper support . . . Harry can give us better success in the future than we have had in the past." A list of names followed, many of whom were workers new to the company and the team. Winning was not the only thing, but it was important to the company and the workers. Sports brought the worker and the Disston family together in a mutual effort to build a common source of pride, which in turn built success.[22]

Moreover, sports elevated the factory to center stage in community life. To be against the company was to be against the community. Sports and athletic associations diverted attention from issues of wage and working conditions in the factory and popularized the company's name in small communities like Tacony. By providing recreation and entertainment for the workers and their families, they also helped improve mental and physical health. Winning soccer and baseball games added to the community's pride in the company, and to the worker's pride in being a part of a "winning team." It was simply good business to stage these events and games under the company's aegis. Despite the influx of new industrial firms into Tacony after 1890, the town continued to be controlled by the central agenda of the Disston Company.

William Dunlap Disston founded another organization with the same intent but a different focus. In a mass meeting held at Henry Disston Public School on Friday evening, February 14, 1919, to "ascertain whether or not the citizens desired a Community Center," he called for improvements in the town's "community life." Guest F. B. Barnes, a Philadelphia representative to the Federal Sanitation Commission, used a slide show to illustrate living conditions in other cities. The new community center recommended by William D. Disston would provide a place for family-oriented activities, such as a domestic science room, where girls could learn sewing and cooking, a music room, a gymnasium for basketball, a pool and billiards area, a game room for checkers and chess, and facilities for debating societies, amateur theatricals, and dancing.[23]

One thousand people had turned out for the meeting, and a constitution committee (half women and half men) was chosen. At the next meeting, on Friday, March 7, the new Community Association heard Powell Evans speak on changing the charter of the city of Philadelphia. Two weeks later, Dr. Susan Kingsbury spoke on community activities. The message to Taconyites was clear: "Come out to the next meeting and get acquainted with your neighbors," said the advertisement in the "*Bits.*" Clearly the assumption of an earlier era that all Taconyites were bonded emotionally and geographically was no longer tenable.

Augmenting and supplementing private community efforts to enlighten the community, restore family values, and control local boys and girls were the public library on Torresdale Avenue, opened in 1906, and the Disston Recreation Center, built in 1912. The center was the first of its kind in the Northeast, and the ground for both it and the library was donated by the Disston family.[24]

Photographs of a play day held at the Recreation Center in 1916 are revealing. The children were engaged in activities that depict the new world of industry. Play was organized and done cooperatively and was less physically brutal than traditional games. Group exercises, the maypole dance, and races helped children develop into healthy, happy, right-thinking citizens. Discipline, order, regimentation, cooperation, and attention to group values were believed necessary for molding future industrial workers. Industrial welfare in the community conditioned more dutiful workers for the future.[25]

Further evidence of the desire to inculcate small-town, family values can be found in other articles in the *Disston "Bits."* In a sequence of stories about workers' families, the editor glorified the large nuclear family. In January 1919 the paper announced a column that would be devoted to Disston men with "representative" (that is, ideal) families with many children. The first family honored was that of Doc Savage, a twenty-nine-year worker at Disston. A picture of Savage, his wife, and their seven children showed what a family should be: large, well groomed, and healthy. "Doc is [a] steady, industrious and skilled workman," the paper said, "and we must compliment him upon his large fine family of which he is justly proud." Al Wellens's family was next: thirty-one years at Disston, Wellens had a wife and seven children. A foreman of the Handsaw Finishing Department, he was described as "skillful and efficient." A month later, John Quinn's family was pictured, and he was described as a worker of twenty-eight years in the Small Circular Saw Department. Then came Jack Hepp, Thomas G. Woodfield, Fred Miller, William P. Norbeck, and Lawrence Donohue. Each had a wife who was a homemaker, at least seven children, and more than fifteen years' service at Disston. The Donohue family received special recognition because their collective service at Disston totaled eighty years. One might add that the continuing labor shortage at Disston could be solved in part by workers "recycling workers"—that is, having large families of which the sons would be led to the plant's hereditary skilled jobs. In this manner some hard-to-fill positions could be filled with reliable workers without having to rely on haphazard recruiting. The nuclear family was essential to the Disston company's maintenance of a skilled labor force.[26]

The concept of the nuclear family promoted specific roles for women, who were expected to leave work when they married. This was especially true of female schoolteachers, who in the 1920s were required to leave their positions at the Henry and Mary Disston Schools upon marriage. In the 1920s the lives of most women in Tacony varied little from the homemaker stereotype. Ninety-three-year-old Marguerite Dorsey Farley remembers her childhood vividly. The houses had no electricity before 1920, no indoor plumbing, and no running water. The water lines ran down Longshore Street and did not reach her childhood home on Knorr Street. Outhouses in the rear of the property and one water pipe in the street provided for waste and water.

The center of life was the family, and "joys came within the four walls of the home." Farley's days as a child had been spent helping her mother with the chores. Much like the workers in the Disston factory, the women of the town worked in regimented and clockwork fashion. No one escaped the pervasive culture of a factory town. Monday was washday, Tuesday was devoted to ironing, and Wednesday to baking for the week (children were sent away or told to play outside so as not to disturb the baking processes). Thursday was spent cleaning the second floor, and Friday was set aside to clean the first floor. Saturday was a day apart, the time for Marguerite's father to work in the garden or repair the house. On Sunday after church the entire family would walk together to Frankford Avenue and take the trolley to Point Pleasant on the Delaware River for a picnic. More exciting were the winters, when they could walk on a frozen Delaware River amid ice-sailing boats flying up and down the river. The fastest of these belonged to tavern owner Charlie Merz. After 1925 the river ceased to freeze solidly due to industrial waste buildup in the river, thus ending this activity.

Chores within the house were done manually. Cleaning before the vacuum cleaner meant taking the rugs outdoors, placing them over a clothesline, and beating them. The floors, wood bases, and sills were all wiped with a damp cloth. It was generally believed that good housekeeping promoted good health for the family. Marguerite remembers most groceries being delivered to the house. The Achuff Tea House delivered coffee and tea once a week, Darreff Food Store delivered food orders twice a week, and Kett Ice Company delivered ice three times a week. Wagons were the most popular toy for the neighborhood's children because they could double as a means of carrying groceries or ice. A produce vender came down the street unannounced three or four times a week. Marguerite's mother had a charge account at Rubin Brothers clothing and semi-department store on Longshore Street. This was the life-style of small-town America at the turn of the century.

Every day Marguerite's father went to work at Disston, never missing a shift. His role was clear: He was the breadwinner and sole support of the family. Her two brothers made their own special contributions to Tacony. Harry won fame on Disston's track and baseball teams. Both worked at Disston, but Frank later entered politics. He was elected congressman for the district in 1934 and 1936. The family remembers with pride Frank's casting the deciding vote for the Tennessee Valley Authority bill. Frank was home for his mother's funeral when he got a call from the White House that his vote was needed for passage of the bill. He immediately returned to Washington to cast the crucial vote.

Marguerite went to the Henry Disston Public School until 1908, when she entered fifth grade at the newly opened St. Leo's Catholic School. The school was so small that the fifth, sixth, seventh, and eighth grades were in one room. From St. Leo's the girls went to Hallahan Catholic High School in Center City, where for two years they took courses in religion, typing, bookkeeping, Latin, English, and history.

The commercial part of town on Longshore Street had shops for every need: wallpaper, plumbing, baked goods, hardware, clothing, ice cream, and linens. The atmosphere in the community encouraged face-to-face trust between store and customer. Newcomers to the community were easily recognizable and had to prove their merit. When Erben-Harding Woollen Mill came to the community, Marguerite Farley recalls, a large number of people moved into the area so they could get work at the textile mill. The Gillinder Glass Company provided a similar impetus for migration of families. Farley remembers five boys from the Toner family and other children from the Richter family who became schoolmates at St. Leo's because their fathers had gotten jobs at Gillinders in Tacony.[27]

The Disston family remained influential in promoting the welfare of the community throughout the years between 1900 and 1929. The Mary Disston School opened in 1900 to relieve overcrowding at the Henry Disston School. Like the library on Torresdale Avenue and the Disston Recreation Center, its construction was facilitated by the Disstons' donation of the land. All three facilities advanced community-company-worker interaction and interdependency, important parts of an organized industrial welfare system.[28]

Disston promoted the welfare of the community in yet another area. Industrial waste was not a concern for most nineteenth-century factory owners, who simply dumped industrial by-products in the river or the countryside, but it posed an expensive dilemma at Disston saw works. Tons of furnace ash were produced each month at the Laurel Street factory alone, and it was expensive for the company to ship it out of the city. Ingenious solutions were devised. Excess ash from the Laurel Street plant was sent on barges to Tacony, where it could be used to cover the muddy roads around the Tacony plant and yards, and metal scraps from saw cuttings were returned to Laurel for reprocessing. Another waste product left from the grinding of saws was a growing stockpile of used sandstones. When placed in service, grindstones were circular and seven feet in diameter, but after two months of grinding they were reduced to four feet. The outer rim of the grindstone now circled too slowly to taper the saw blades. The firm paid ten workers full salaries to move and install the large grinding stones. It was too expensive to ship them away, so the stones were cut in half and used to build walls around the Disston Recreation Center, a river wall, and parts of the factory. At the request of the Baptist church on Disston Street, the company donated used grindstones for a church that cost the congregation a mere $11,000 to build. Foundations for streets in the community were also layered with sandstones. Sandstones remain as prevalent in Tacony today as they were a hundred years ago.[29]

Efforts that united the community at a time when ethnic diversity was threatening to divide it were important to Henry Disston & Sons. Libraries, schools, recreation centers, and fraternal organizations, all supported by the company, promoted discipline and order in a community that was receiving new immigrant groups. The events of World War I furthered unity in the Tacony community and loyalty to Henry Disston & Sons.

As had been the case during the Civil War, Henry Disston & Sons' steel-making capacity was an important national resource during World War I. Production for the armed forces between 1917 and 1918 consisted of rifle gun-barrel steel, bayonets, light armor plate for tanks, and gun shields for the navy. Increased production meant a larger workforce and modernized processes for steel-making. It was evident that the steel-melting capacity used for saws and files was inadequate for the demands of wartime production. President Frank Disston (1915–1929) decided to replace the old crucible pots with two new melting furnaces, one with a three-ton capacity and the other with a six-ton capacity. A six-ton steel hammer was placed in the same building for hammering the larger ingots poured from the new furnaces. (The war ended before the new furnaces were installed, and peacetime brought a decline in the demand for steel, so that the plant did not operate until 1925.) The armed forces' need for men decreased the working population available to Disston. Blacks, Italians, and Poles were now welcomed to the firm to fill the void of Disston men serving their country in France. War also forced the community to come to grips with its larger and more diverse population.[30]

Making steel for war purposes brought the company and Tacony together in a common effort. Cooperation increased, and workers did their part for the boys overseas. As the nation's attention became riveted on World War I, so did Tacony's. The Pennsylvania National Guard unit was the first group to leave Tacony in preparation for war. Their leader was Dr. Elmer E. Keiser, the organizer of the unit some years before.

Keiser had graduated from the University of Pennsylvania Medical School in 1890 and established a practice in Tacony in 1906 after having been appointed the Holmesburg Prison doctor. Known as Company M, First Regiment of the State Guard, Major Keiser's Tacony boys first served the country at the Mexican border. When World War I began, the unit was sent to Camp Hancock in Augusta, Georgia, as the 110 Field Hospital Regiment. Disston men made up the majority of Tacony residents going off to war.[31]

No sooner had the troops reached Georgia than Edmond Roberts, secretary of Henry Disston & Sons, learned that the Tacony boys did not have saws to cut firewood. He rushed an order of saws, free of charge, to his "community away from home." Paul Biemuller, son of the local Lutheran minister and a former Disston employee, reported that the saws were "distributed amongst the 'Disston Boys' immediately on receipt. They all had to try them right away; were so glad that at last they had a decent saw to cut firewood with. Have been using a hand saw up to now, cutting logs 6 and 8 inches in diameter and it sure is a pleasure to use a regular, I mean 'The only saw on the market.'"[32]

Tacony's small-town values and sense of community are illustrated by one of the town's best-known war stories. When the Tacony corps was to go overseas to France, one of the men phoned to alert everyone in Tacony that the train taking them to New York would travel through the Tacony station. Engineer John Costigan, a resident of Tacony, stopped the train at that station despite orders to the contrary from the military. Costigan wired his superiors that the train had stopped because a "hot box" required repair. The whole town was alerted, and men, women, and children in night clothing headed for the station for a last good-bye to their loved ones. Privately, Costigan admitted, "I couldn't come back to Tacony and face the people if I went through to New York knowing I had the boys from Tacony on the train. Further, my mother would disown me. I would never think about not stopping the train."[33]

There was deep concern for the welfare of boys fighting in France. The whole town knew when letters arrived from the corps, and the local mailman was known to tell the children on their way to St. Leo's or the Disston school about letters from brothers overseas. Marguerite Dorsey Farley remembers vividly being told by the postman of a letter from her brother, Harry Dorsey, in France. She was happy all day in school even though she had to wait until she got home to read the letter.[34]

Tacony proved, predictably, to be a very patriotic community. Letters from Tacony boys were printed in the *Tacony New Age* and the *Disston "Bits."* The stories were of brave sons fighting for their country. Mrs. Fisher, Mrs. Brady, Mrs. Dorsey, Mrs. McMenamin, and other mothers met at Disston Playground once a month to read letters from their sons to one another.

The community was stirred on August 17, 1918, when the *Public Ledger* reported that two Tacony boys were the heroes of an encounter at the front in France. On August 9 and 10, Mike Biemuller, son of Tacony's Lutheran minister, and Albert Baker evacuated twenty-eight men from a battlefield at Fismette. Ten of the thirteen ambulances of their company had been destroyed by shellfire. Under fire and with complete disregard for their own personal safety, they located a line of evacuation to the rear for ambulances. With just three ambulances, the Tacony boys, along with another Philadelphian, labored for forty-eight hours, driving through a shell-swept and gas-infested area to evacuate wounded soldiers.[35]

Soon after the *Public Ledger* was received, crowds with American flags and banners formed outside the two soldiers' residences. Mrs. Anna Baker of 3027 Longshore Street greeted her friends in front of her house, telling them, "It is worth every hour I put in for this truck farm to read anything. It's just great. I'm an American and I feel just so proud of that boy of mine that I feel foolish. Look at those flags out there and every one of them is out for my boy." The Rev. Andreas Biemuller of the German Lutheran Church at 6812 Jackson Street shouted the news to his wife as he greeted the crowd outside their house. Biemuller, with justifiable pride, went all over Tacony letting everybody know that his son had driven one of the ambulances that evacuated the wounded Americans. Later, when Mike was presented with the Distinguished Service Cross by General Pershing, the whole town attended special services held by Reverend Biemuller.[36]

Parades and flag-raising ceremonies were common in the community. On Friday, August 2, 1918, a Disston worker named Hanlon was the main speaker at such an occasion, just a few days after he had learned that his son was in a hospital in France suffering from gas poisoning. Hanlon's speech conveys the level of patriotism felt by the people of the town:

> Let us at this time and in few words render homage and devotion to a flag that represents the highest principles that man has ever contended for:
>
> Oh! sacred emblem of a grateful people
> We meet here to do thee honor,
> Living beneath thee as the guardian of our liberty
> We here render testimony to the beneficence of thy work.[37]

This need to support the boys at the front, along with calls for wage-earners to support the government through Liberty Bonds, was proclaimed by the Disston company throughout the war. The newly constructed movie house on Longshore Street was named "The Liberty" in honor of those who fought in the war. The Tacony Flag Association planned Memorial Day parades and other celebrations with the company's support. The largest of these parades took place on July 22, 1918, when William D. Disston led the workers from the factory at noon in a spontaneous outburst of loyalty and support for the boys in Europe. In an orderly march, two thousand workers, men and women, carried flags and stepped to the music of the factory band up Longshore Street to Torresdale Avenue and then back down Disston Street to the factory, where they returned to work "with renewed vigor." A "Liberty Sing" was organized by residents of Van Dyke Street between Unruh and McGee, a block that had seven boys serving in the war. Similar celebrations were held in the community during World War II and the Korean War. It was not until the Vietnam War that the community questioned sending their boys off to war.[38]

The cry "Your country's job is yours" echoed throughout the plant and prompted poems in the *Disston "Bits"* expressing the workers' patriotism:

> Some people read our Magazine and others do not,
> But I'll tell you about the File Shop boys, who handle things red hot.
> We are working for our Country for this inhuman war to win
> And we are backing up our heroes with all of our spare tin,
> For we helped to fill up the War chest to one hundred per cent.
> And we all have got a Liberty Bond—to our Government we have lent. . . .
> And will help to Lick the Kaiser, who doesn't give a damn,
> For all our Yankee heroes and our dear old Uncle Sam.[39]

In all, 314 Disston men served in the armed forces during the war. Only 18 were killed. The first man in the community to fall was a nineteen-year-old named William Oxley, who had enlisted in the infantry rather than the 110th Ambulance Corps. (No soldier from the Ambulance Corps was killed.) The returning veterans named their American Legion post after the fallen Oxley.[40]

The year 1918 proved to be the critical one in redefining Tacony's contribution to the war effort. Activities at Disston & Sons reflected the national transition from domestic to wartime economy. The year began with the company in the midst of a shutdown due to a lack of coal, which lasted from December 14, 1917, to January 14, 1918. The result was a decrease in production and two hundred layoffs. Unable to regain these workers, Disston looked to blacks and women to provide a substitute labor force. This triggered the first wave of black hirings, and the number of women workers increased correspondingly.[41]

Both groups were assigned to entry-level positions at the factory, and white workers in those positions were upgraded to more-skilled jobs. Blacks became laborers in the steel works, where sheer strength was the main qualification, and women, mostly young and unmarried, sanded and shellacked handles in the saw and the packaging departments. The motive behind the work placements seemed clear. Blacks carrying steel in labor gangs were not working side-by-side or on an equal basis with whites, and women were quartered in closed departments where management felt that their presence would not upset the work routine of the men. Such women came from the Italian and poor Irish neighborhoods of Tacony and the Polish neighborhood of Bridesburg. Middle-class Tacony women never worked at the factory, even during the war years, preferring to donate their time to Red Cross sewing activities on behalf of the troops. Work at Disston during wartime was done out of economic need, not patriotism. However, such workplace changes promoted company consciousness that blacks and women could be and were useful workers. When the war ended, some of the best of these workers stayed on in laboring positions, and small pockets of female and black workers have remained part of the Disston workforce until the present day.[42]

The company needed to increase the output as well as the quantity of workers. A new production department was organized, and the Employment Department was directed to scrutinize the quality of men and women hired and to be sure they were placed in the positions most commensurate with their abilities. The foremen were given a dinner at the City Club to underscore the necessity of continually upgrading the quality of the product while increasing quantity and reducing costs. Mass patriotic workers' rallies were held every two weeks, and speakers from the War Department talked to the men. The War Department also allowed Disston to utilize the services of Antoinette Greely, a federal government employee on loan, to establish better rest-room facilities for women and increase the lunchroom amenities. Through Greely's efforts, a cafeteria was established for workers in 1919.[43]

A Department of Intensive Training was organized, giving the more expert workers responsibility for teaching less experienced ones. Thus, the foremen were relieved of a measure of their normal training responsibility, giving them more time to monitor the routing of work and keep the machinery in good order, and avoiding shutdowns on the production lines.

In addition, the traditional workday was restructured. At the firm's founding, Henry Disston had established a work incentive to entice new employees. Workers were free to leave the plant in the afternoon once they had met their work quota for the day. But from management's point of view, this was a great waste of productive labor time, because the plant workers averaged only seven and three-quarter hours a day. To increase war production, workers were required to serve a full ten-hour day. After March 1918, wartime plans restricted workers from leaving the plant unless they had a pass from the foreman. Disston & Sons reduced "wasted" employee hours by 660 a day by the simple expedient of eliminating early departures.[44]

The end of the war evoked special commendations from company vice-president William D. Disston for the workers' "splendid support and loyalty to their country through that never-to-be-forgotten, hard, but glorious year of 1918. I wish to thank the superintendents and foremen for that support." To William Disston it was obvious that cooperation and teamwork, the essence of his managerial style, were what made the nation and the company so successful during the war years. He reminded the workers, much as his grandfather Henry had done years before: "We are all associated here together in one business and when this business is prosperous we are all prosperous." He continued that the company's motto for 1919 should be "High Quality and Reasonable Costs Bring Orders and With Them a Full Dinner Pail." After the speech, the workers gave a rising vote of thanks and three rousing cheers for William D. Disston and the company.[45]

When the war ended, most of the men in the 110th Ambulance Corps were retained in Europe to look after the American wounded who were too ill to be sent home. Living with French families during these months, they were among the last Americans to leave Europe. On their arrival home, the rejoicing community held a parade, dances, and celebrations to honor the boys from the corps. Bessie Hicks, who had organized the Ladies Emergency Aid during the war, now organized formal dances and community celebrations. Marguerite Dorsey Farley remembered these celebrations but was surprised that few of the men spoke openly of their days in Europe. Nevertheless, at community get-togethers the same veterans could be found in private conversation with their buddies discussing war experiences.[46]

The patriotism exhibited by Taconyites during World War I continued through World War II, the Korean War, and the Vietnam War. Such manufacturing towns had a great stake in wars. Disston's steel plants were always awarded large war contracts, which benefited the community with increased work and enhanced prosperity. The success of World War I, and the relatively small number of soldiers killed, also helped promote patriotism in Tacony, which had prospered greatly while remaining relatively unscathed. A festive, fraternal pride colored their recollections.

For Disston & Sons, as we have seen, the war meant increased orders and profit. Demands for steel for shells, rifles, and military materials meant more work than the community could handle. Women and blacks were beneficiaries of an expanding workforce. In an open letter entitled "Teamwork Always Wins," S. Horace Disston, then a vice-president of the firm, made clear his success formula soon after the war ended. Disston reminded the workers: "Everything in the way of words, actions or deeds that gains credit for Disston reflects credit to you." Quoting from Rudyard Kipling, Disston let the workers know how they could best serve the company.

> It ain't the individual
> Or the Army as a whole
> But the everlasting team work
> Of every blooming soul.[47]

Henry Disston & Sons enlarged the variety of industrial welfare techniques in an effort to deal more effectively with the workforce after the war. Within the plant itself, these new programs took four forms. A new Personnel Office was opened, with returning war veteran Harry Dorsey in charge. Dorsey was a graduate of the University of Pennsylvania and a son of one of the men who helped open the factory in Tacony. The mission of the office was to interview all applicants, to control the flow of workers by matching applicants with jobs requiring their particular skills, to survey the shops to determine each one's specific labor needs, to promote efficiency programs to help workers adjust to the work, and to conduct exit interviews with everyone who left the company to determine and evaluate the reason for departure.

The personnel office was responsive to a number of pressing problems. Originally, one got a job by seeing a foreman. Favoritism and subjectivity at this level sometimes led to the employment of workers who were inadequate to their current position or not ready for a more skilled one. Central office hiring would put some limits on this practice, although it is clear from interviews with former workers that many persisted in bringing notes of reference from foremen or higher-ranking department heads to the personnel office.[48]

Although the company's first priority during wartime and afterward was solving the shortage of labor, it was hoped that the Personnel Office might simultaneously enhance workers' welfare. The exit interview was an important part of this strategy. If a worker was leaving because of dislike for a specific job, the personnel office might be able to save that worker for the company by reassigning him or her to another type of work. This was especially true of the unskilled laborers, who appeared to drift in and out of the workforce daily. Cutting down on turnover was seen as an important method of increasing production, since the company was currently wasting too much time training new laborers.[49]

Worker loyalty was fostered by other measures. The cafeteria provided nutritious hot meals for lunch at a price lower than many restaurants in Philadelphia. Hugh Moore claimed to have had "a dandy dinner there today for 35 cents. It's great. I paid 65 cents at another place, and it wasn't any better. The cook can't be beat." Mrs. Frederick W. Melsch was even more laudatory:

> I think [the cafeteria] is the finest thing Disston ever did for their men. . . . What a relief it is to realize that I do not have to plan for that lunch every day, and arise a half hour earlier every morning to prepare it. Then the change of environment while my husband is eating his lunch is beneficial. I know he will feel better to get the fresh air going to and from the Cafeteria, mingle with his fellows. . . . A work shop was never intended for a dining room and I for one think Cafeteria supplies a long-felt need.[50]

Women workers applauded their new rest room instituted during the war but retained because of its positive effects. It contained comfortable chairs in a lounge area enhanced by curtains and other decorations. The bathroom, in a separate but adjoining room, was new and exceptionally clean.

The company's interest in industrial welfare programs prompted William Disston to ask Dorsey to investigate a pension plan for employees, and by 1920 the rudiments of this plan were in operation. It contained provisions for a maximum retirement benefit of $100 a month for an employee after the age of seventy. Despite its obvious limitations, it was welcomed by the workers as further indication of the company's interest in their welfare. Pensions were aimed at tying employees to the company for life and represented to the company an inexpensive way to deal with worker turnover and training.[51]

Nevertheless, President Frank Disston faced a myriad of problems in the difficult transition back to peacetime production. Profits in the early

1920s were dismal, and new strategies were needed if the company was to pay for these new benefits. First, the crucible stages were torn down and steel for all products was melted in the new three-ton electric furnace constructed during World War I. Next, the saw-handle shop was reorganized and a new lacquer tank that eliminated all hand-painting for handles was installed. In the main hardening shop a continuous-feed hardening furnace for hand-saw sheets was designed and built, cutting direct labor costs and improving quality.

A special research project to develop a light-gauge armor plate was begun at the plant by Henry Allen, head of the Disston metallurgical department. Allen, later president of the Franklin Institute, began with the German World War I armor steel mix and improved its resistance strength and lightened its weight. By the 1930s these refinements produced a formula for armor plate that was the finest in the world.[52]

Yet another new product begun during Frank Disston's reign was the sugar beet knife. After World War I the United States turned to raising sugar beets to supplement the importation of sugarcane. A machine for shredding sugar beets was produced by the Ogden Iron Works in Utah, whose officers approached Henry Disston & Sons to make the knives for their machine. Disston's chief engineer, Norman Bye, and master mechanic, Sam Freas, used a steel formula that called for 1,060 carbon steel treated to a 46-49 Rockwell C scale. The steel was purchased from Bethlehem Steel Company rolled to shape but machined and heat-treated at the Disston plant. Bye and Freas followed up on this invention (see Chapter 7), allowing Disston to enjoy a virtual monopoly in the beet-knife business for the next twenty years.[53]

Paternalism had not completely eroded. There is no evidence that Disston's workers were dissatisfied with the company, despite the growing ethnic pluralism of the workforce. Nor was there much criticism of the Disston family, whose benevolence and gifts of land were widely noted.

For this period between the wars, Lizabeth Cohen describes an American society where workers were becoming more unified by means of national radio shows, movies, and chain stores. Unity in Tacony had been directed and orchestrated by the company, which sought the best workers at the least expense to itself and was aimed at keeping diverse groups working together. As workers began to think as one, their demands became more focused. Interest now centered on individual programs, not on community programs. Also, the Great Depression provided the dynamics that encouraged the shift to an interest in job security, higher wages, and health benefits. Unionism became the mechanism needed to force reluctant companies to accept these unified demands from a workforce that was frightened and confused by the depression. At Disston, unionism eventually ended paternalism and lessened interest in industrial welfare as well.[54]

CHAPTER

6

Inside the Factory in the 1920s

Plant Design, Worker Skills, Turnover, Sales, and Quality

IN THE RELATIVELY TRANQUIL TIMES, 1920s Disston workers began the day shift at 7:00 A.M. by punching a time clock at the gate. Skilled workers picked up the previous day's work; laborers reported to the foreman for an assignment. At 9:00 A.M. everyone in the factory had a half-hour break. Workers learned to time their work so the machines could be shut down for the break. At 9:30 A.M. it was back to work until noon. Lunch lasted forty-five minutes, and a straight stretch between 12:45 and 4:00 P.M. finished the day. The regimentation resembled what most of the workers had experienced in the public schools, but with a foreman in place of the teacher. This system worked well because the skilled workers knew their craft and the foremen were collegial in their relations with them. The laborers were there to support the skilled workers and the production process. Laborers who were willing workers and kept their eyes open as they moved about the plant might be promoted to machine operators or even to skilled smither or smelter. The factory was a land of opportunity for the newly hired.[1]

Throughout these years, much of the work was done in what today would be considered unhealthy conditions. Given the knowledge of the time and the available remedies, the Disston plant had an enviable health and safety record. Yet accidents were common: burns, cuts, and subsequent infections, as well as back and muscle injuries due to lifting. Moreover, the air in the grinding and steel factories was filled with sand and small steel particles. Wherever saws were used for cutting the finished product into lengths, there was steel dust. Many a worker justified a daily glass of whiskey at Merz's or Halbert's taverns on the grounds that it "takes the dust out of the throat." Long exposure to such an environment could result in some form of lung disorder. The steel factory and rolling mills were suffocatingly hot in the summer, and moving back and forth between the cold outside and the heat inside left workers with continuous "colds" during the winter. Despite these hardships, the Disston men interviewed were not upset with their treatment: "It was the way things were and we adjusted."[2]

The factory office was the hub for all plant activity. About one hundred office workers handled factory orders, payroll, hiring, and secretarial work. Communication between departments was provided by ten boys between fourteen and sixteen, called "runners." They would sit in the main office, and when a card was flashed with their number on it, they rushed to the desk, picked up a message, and were off to the specified department foreman. This system of communication prevailed until the advent of a mechanized sound system that could blast information through loudspeakers located throughout the plant. The foreman received orders for a specific number of saws of a specific size and shape. He then reinterpreted that order into work after determining the number of such saws in storage and the general trend of orders. There was a lot of guesswork in such estimates of future demands. Nevertheless, the factory operated under this system until the company was sold by Disston in 1955.[3]

There was one major effort to streamline the management of the plant after World War I, when tensions between management and labor came to focus on production issues. The Disston company's solution to the problem contradicted the commonly held beliefs of the time about industrial management, which centered on the scientific management principles of Frederick W. Taylor. Taylor's emphasis on making manufacturing more efficient through measurements was never used at Disston. Skilled workers in saw and steel were bound to their craft and looked askance at suggestions from those not in the field. As late as the 1950s, skilled worker William Rowen resisted what he considered the ill-informed advice of a young efficiency expert sent to the factory by the new owners, H. K. Porter.[4]

The Disston plan rejected expert problem-solvers and scientific experimentation and instead chose group interaction as a way to manage the plant. Disston's managers recognized that participation of worker-foremen was essential to the improvement of production. Coordination, cooperation, and recognition of how individual departments fit into plant operations as a whole were the basis for continual improvement. Under this plan the various department heads would have information about the working of the whole firm and be consulted on actions and decisions before implementation.

The plan grew out of William D. Disston's suggestion in 1919 that a special committee be appointed to oversee the interrelationship of departments and the subsequent effects on product quality and production. The departmental system retained the direct line for orders from department heads to foremen. To add the cohesion and flexibility necessary to maximize production, William Disston added supportive departments. This new line/staff arrangement was aimed at eliminating wasted worker hours. With quality control rigorously assigned to each department, a new interrelationship of departments held the key to speeding up production.[5]

William Disston's war experience gave him this vision of how the company should be reorganized. He became head of the Production Department in 1917, a position that in his judgment allowed the best understanding of the daily problems of the factory. A shortage of steel in 1918 affected production in the saw, tool, and file shops and called attention to the breakdown of communication between shops. In wartime, steel earmarked for armor plate and bullets had been misdirected to the domestic shops. The system of shops malfunctioned; it lacked coordination.[6]

So it was that William D. Disston convened the first meeting of what became the General Factory Committee, which consisted of key workers from the various departments in the plant. Under Disston's plan, the former department foremen continued to serve as monitors for product quality and training of the skilled workforce, but there would be eight interdepartmental superintendents in a new production department. Their responsibilities would cut across what were formerly shop responsibilities. The responsibilities of these superintendents as stated in their titles convey much of the new organizational plan: Workmanship, Heat Treatment, Metallurgy, Equipment, Planning, Costs, Routing, and Materials and Labor. Clearly, production was the dominant theme of this new system.[7]

The Factory Committee was to act as a cross-department problem-solving agent and an adviser to management. The main concerns of this committee were the shortage of workers and the quality of the product. Most discussion focused on what committee members called the "labor turnover" issue. To understand the committee's thought process at the time, one must know a little about the various job levels in the plant. There were five factory-level positions based on skill and salary. In order of pay and importance to the company, these were (1) the foremen; (2) the skilled laborers, the backbone of a steel/saw factory; (3) the millwrights, machinists, and mechanics, individually trained in the wood and steel trade to make parts for products and to keep the plant machinery operational; (4) the machine operators, crane operators, and workers who ran production machines; and (5) the laborers, the unskilled men, boys, and women of the plant. Training for each group varied. The highest training was required of foremen, because they had to be the most competent workers in their trade. It was not unusual for a skilled worker to wait twenty years for promotion to foreman. At one point William D. Disston once suggested merit pay for foremen, believing that such a bonus would stimulate production. The criteria would be the quality and quantity of production, the method of paying employees, labor turnover, the attendance record of employees, interest in operating methods (willingness to learn), and years in service. Heads of departments disagreed, pointing out that such policies led to jealousies among workers and were not productive. The foreman was the key to getting work out, but he was not to be put in the position of having to push production quotas. Disston backed off, but he did insist that foremen be rated in these areas and their rankings posted in the shops throughout the factory.[8]

Smithers and smelters served a Disston apprenticeship for four years at a laborer's pay. The mechanic or skilled metal machinist was required to serve a similar apprenticeship period while earning a laborer's hourly rate. Machine operators were given thirty days to learn the machine, as were crane operators, and trainees were paid the laborer's rate. Laborers were hired one day and started work the next as part of a group of four to six men. A laborer did what the group did and followed directions, asking as few questions as possible lest he be considered a troublemaker. Skilled workers trained and ready to work were difficult to find. Smithers and smelters were brought from England and encouraged to live in Tacony and have children who could be recruited through a preference policy that allowed them to follow their father's trade. In addition, laborers who showed promise could apply for and gain entrance into a better job.[9]

Russell McIntyre began working as an unskilled laborer, trucking material from one department to another, in 1941. After a few years there was an opening for a machine operator in the file shop. McIntyre applied, and his good work record got him the position. He was trained for thirty days by another machine operator, and then he was on his own. As a file machine operator, he ran four machines simultaneously, each machine producing a file every ten minutes. McIntyre's machines cut rounded and flat files. If a machine went down, the foreman was told, and the machine operator received the minimum hourly pay for that machine while it was being repaired. Broken machines usually meant short pay that week. Disston required the men to make thirteen files, a baker's dozen, to receive piecework credit for twelve.[10]

When the factory was on shift work, the machine operators usually looked out for the man who followed them on the machine. During the war, McIntyre would always run an extra few dozen files for the next worker, to make up for time lost in restarting the machine. When the factory was on three shifts, McIntyre could expect to find a few dozen files waiting for him each morning. Workers did not misuse their machines during their shifts. Repairs were made promptly—supplies and file blanks were kept on hand. This cooperation between workers on the same machine enabled the system to work. Discipline, order, and cooperation were the tenets of a good factory worker. A functioning, well-maintained machine gave all three workers on shift work a chance for good money.[11]

When the file shop was closed by H. K. Porter in 1956 (see Chapter 9), McIntyre went to work in the saw-grinding department, where he was paid a laborer's basic wage for three months until he learned the grinding trade. Despite maintaining his plant seniority, McIntyre lost his job seniority and was paid a lower rate because of the union rule requiring that job seniority begin anew for those who changed shops.

The mechanics' shop was the in-house repair shop. Built especially to devise and repair machinery, it was completely equipped with the latest drills, cutters, and grinders before it opened in 1909. Connected with the shop were the blacksmith shop and the brass foundry. All three shops were supposed to do repairs quickly and keep the machines of the factory running. Most of the machinery was old, and no spare parts were available. Using high-precision tools and cutters, the machine shop duplicated parts for nonfunctioning machinery, allowing Disston & Sons to use nineteenth-century grinders, teeth cutters, and steel-cutting machines into the 1950s. Former Disston workers remember the patchwork of their machines. Few machines on the floor were without a Disston-built replacement part. In the days of cheap labor with no social security or retirement and medical plans, a company could afford to devote a shop to making parts for machinery. Today, when personnel is the major expense for most companies, it is cheaper to replace equipment than to repair it.

Disston & Sons had other reasons for keeping antiquated equipment. Even in periods of economic prosperity, getting new machinery would consume managerial time, disrupt production facilities and schedules, and might cause labor conflict within the factory. It was simply easier and more pleasant for everyone to declare dividends and give the excess money to the family. Under these circumstances, the Disston Company rarely set money aside for new equipment, a condition that resulted in a plant that needed extensive modernization after World War II.[12]

The large unskilled labor force was also possible because of the minimal impact it had on the economics of the plant. As noted, there were no benefits or medical payments, and the pay rate for boys was as low as $4.50 a week in the 1920s and 1930s. Laborers were responsible for moving steel in and out of the mill, either manually or with the help of cranes and machines. There were laborers to clean and maintain shops; messengers for communication; and helpers for smithers, machinists, and smelters. Transit crews could be found in the lumberyard and sandstone departments. Work was hard and pay was low, but these laboring positions were the entry-level positions for most factory workers.[13]

Disston's skilled workers were the heart of the plant. Sam Freas was the head mechanic in the chisel-point shop (making removable and easily replaceable teeth for circular saws, thus eliminating the need for resharpening) in the 1920s and 1930s. Sam's responsibility was to keep all the machinery finely tuned in order to produce the best-quality saws. Through the use of precision machinery, and his own knowledge of machinery, steel, and the demands of the customers, he was able to reproduce any piece of steel to accurate scale. His assistant was Albert Gehr. They both kept detailed descriptions of how to make specific products in a pocket-size "black book" they carried with them each workday. Most of the skilled workers at Disston kept such books.[14]

Albert Gehr's book remains. It is a remarkable description of how one worker recorded information he thought essential to his part in the making of inserted-tooth circular saws in the years between 1924 and 1927. Gehr was a quiet but highly respected man, and his writings show the critical thinking skills, logic, individual responsibility, and memory necessary to succeed as a machine operator in that day. The book contains kerf (width between teeth) measurements transferred into machine-gauge measurements. Gehr also kept records specifying the side clearance of inserted tooth saws that gave precise measurements so the final product could compensate for the expansion in the running process. This allowed the teeth to bind in an immovable position and still not buckle the steel plate of the circular saw at high speeds. There were also warnings to check the width of circular saws to determine the size of the teeth to be used. The slightest variation from this exacting procedure could ruin the saw. Specific orders for companies with unique steel requirements were noted and described so that steady customers could count on orders being filled in accordance with their needs. Notes were added as deemed appropriate.[15]

The book suggests that factory workers were in many cases far more independent and critical in their thinking than is widely thought. Their work was not a matter of making the same product every day, but required mastering the machine and recognizing their part in the final product so the customer would have a product of the desired quality. At the root of the process was worker control over making the product and providing for customer satisfaction.[16]

An examination of the millwrights and brass smelters illustrates Disston & Sons' penchant for quality. Millwrights had a full lumberyard of applewood for making saw handles. The wood had to be seasoned in the open air for three years before it was used. At that point, millwrights used handsaws to cut each handle separately. These were sanded into shape, the final product being a beautiful, hand-crafted handle made of one of the hardest and most durable woods. The brass foundry was established specifically to make brass screws to hold the saw blade to the handle. The keystone logo and name were molded on a screwhead. These finishing touches gave the Disston saw its distinctive look. Again, craftsmanship was the key to this operation.[17]

As electricity came into common usage, large electric cranes replaced steam-operated cranes for bulk-moving heavy loads in the factory. The electric cranes were easier to operate and could be constructed as part of the factory building. The 10-ton Sheppard Niles electric crane was purchased in 1910 and placed in the hammer shop portion of the steel plant. The position of crane operator, with its high pay and lower physical exertion, was much sought after, despite the 75-foot climb to the crane seat.[18]

Bob Bachman started in 1936 in the steel hammer shop. He began as a laborer, picking up steel scraps and putting them in a barrel for 30 cents an hour. After six months he was put in the yard, making 35 cents an hour carting steel ingots on a special truck. A year later, Bachman was manning the furnace at 75 cents an hour with overtime. In another two years he became a gang leader of workers loading the furnace. He was now earning $1 an hour. With each job change, Bachman learned new skills and progressed toward a salary that could support a family. He achieved special status in the factory when he was trained on the Sheppard Niles crane in the hammer shop and was paid $4.50 an hour. He had worked almost twenty years at Disston before he earned that position.

The daily routine of the steel plant depended on the crane operator. In a building that was longer than a football field with no wall barriers, the cranes could swiftly move hot steel from one location to another. The electric furnace poured ingots into molds, which were stripped and carried by crane to the hammer shop. Ingots weighing 600 pounds each would be heated in furnaces and then removed and brought over to the steam hammer. The hammer's 6-ton blows compressed the steel so no air was left to weaken it, and reduced ingots to billet size so they could be sent to the rolling mills to be rolled into sheets and bars. Billets were of two kinds: rod and sheet. Their sizes varied from 50 to 250 pounds depending on the product to be produced. Ineffective use of the crane delayed the work of everyone on the floor, and negligent operation could be dangerous to the workers. The crane operator became the pacesetter for production. It is little wonder that Bachman spoke so proudly of his work and responsibilities some thirty-five years into retirement.[19]

Smithing was a highly skilled trade. Jobs were passed down within families or given to men with connections at Disston & Sons. A smither had the responsibility for honing out the imperfections in the steel, erasing the high points and valleys in steel saw blades, and setting the correct tension in circular saws. Whereas simple flat saws might need little work, smoothing the tension in circular saws always required many hours, because centrifugal forces caused the outer rim to vibrate at high speeds. Tensioning was a skill that took years to master and made smithing one of the most selective and desirable occupations in any saw-making operation.[20]

Bill Rowen was master smither of saws at Disston for more than thirty years. He was born in Tacony in 1911, son of wallpaper-store owner Ellwood Rowen. After leaving school, Rowen began his search for work. He first applied at the School District of Philadelphia for a custodial position. His uncle, William Rowen, was school board president, and he was told to come in for an interview on a Monday morning. Unfortunately, William died over the weekend, and young Bill never got the job. Later the same week, his uncle George Gebhart, a superintendent at Disston, came to a family gathering and, hearing of Rowen's plight, told him to apply for an opening as a smither's apprentice at Disston & Sons. Rowen got the job. In 1928 he began working for expert smither John Southwell. Learning fast from old-timers like John McCormick and Ben Taylor, Rowen soon became one of the more skilled workers in the shop. Southwell became foreman of the smithers and relied heavily on Rowen for difficult jobs. When Southwell retired in 1939, Rowen took the position as foreman of the Smithers' Department and remained there until his retirement in 1978.[21]

Working side by side with the smithers were the holders. John Hansbury was a holder for a short period during World War II. The pay was that of a laborer but the work was much more dangerous. Holders positioned the circular saw flat on the anvil so the smither could use his hammer to make the necessary corrections to the saw. A nonlevel position or an inability to hold the work in position could result in a broken arm or back injury from the vibration of the saw. Hansbury left the Disston job because of low pay and the tedium of the work.[22]

Smelters made the steel. They determined when a melt of iron ore was ready for a measured amount of carbon to be added. The smelter learned the job in years of apprenticeship and involved what appeared to outsiders to be mystical and intuitive skills or even guesswork. In fact, smelters could gauge with amazing accuracy the correct amounts of carbon and the appropriate time for mixing it into the melt. Sam Needham, a longtime boarder in the home of master smelter Jonathan Marsden, remembered when the workers would try to add charcoal to the final heat of the day to speed up the process. Marsden was always nearby to kick someone in the pants and call him a "bugger," because he knew that introducing charcoal to the process too soon weakened the steel. Even more delicate was a process whereby saws were tempered by reheating the steel in vats in the hardening shop. Too much or too little heat could result in a defective saw. These skilled craftsmen were modern-day alchemists.[23]

The original smelter for Disston & Sons was Jonathan Marsden, who was born in England on July 14, 1829. He learned to make crucible steel in England and came to America as a boy seeking employment around 1850. He originally settled in Pittsburgh, but was attracted to Disston & Sons by an offer of advancement and higher wages. In 1855 he opened Disston's first crucible steel factory and became the company's chief smelter. By 1869 Marsden was one of the most valued employees in the firm, as indicated by his substantial salary for the year ($9,421.32). The harder, more flexible steel produced by the crucible process gave Disston a tremendous advantage over its competitors.

An experienced smelter like Marsden determined when to add carbon to a melt instinctively, based on the color and consistency of the iron in the batch. If carbon was added too soon, the steel finished faster but had too much oxygen, making it weak. If it was added too late, the steel did not harden completely, again weakening it. At the time, no smither in America could more accurately determine the exact moment to add carbon to the heat. Marsden's apprentices continued the craft process in Disston's steel mill until the introduction of laboratory testing. In 1883 the first carbon analysis was made at the factory, and by 1898 a complete chemical laboratory for conducting chemical and physical tests on the steel had been installed. The accuracy of the smelters was checked by these new scientific methods. According to Samuel Needham, when it came to judging how much carbon to add to a heat of steel and when, the workers' judgments were as accurate as those after lab tests.[24]

Marsden supplemented the apprenticeship system by going to Sheffield, England, each summer to recruit smelters, smithers, and other experienced steelworkers. The company paid for the trips and provided homes for those Marsden was able to entice to Tacony. During one of these trips, he brought Sam Needham back with him to work as a smelter at Disston & Sons. Needham lived in Marsden's home and worked at Disston until he retired in the 1930s.[25]

The company valued these skilled workers. Men of skill were rarely if ever laid off. During hard times, skilled craftsmen could become laborers at laborers' pay, holding circular saws or moving materials around the shop until increased orders put them back into their skilled position. Therefore, the burden of slow times fell hardest on the least-trained laborers in the factory. However, turnover in these areas was so great that Disston rarely needed to fire a worker. Most retired skilled workers believe that Disston never did so.[26]

Disston management never harried or hassled skilled workers. The normal procedure was that the foreman made up work cards for a job, and the skilled workers listed how long it took them to complete a procedure. William Rowen, a former foreman, remembers well that men adhered to an unwritten but mutually acceptable schedule. Rowen estimates the time needed to complete a 72″ circular saw as follows:

Hammering (smither)	3 hrs
Rough grinding	1 hr
Anvil flattening and tensioning (smither)	2 hrs
Second grinding	30 mins
Polishing	1 hr
Blocking (smither finishing tensioning)	3 hrs
Sharpening (hand filing for large saws)	2 hrs
Etching placed in saw	5 mins
Packaging	15 mins

Thus, it took about 12 hours and 50 minutes to produce one saw, and 9 of those hours were in the smithing process. Foremen did the estimates, which were made possible by the number of experienced workers in the plant.

Workers stayed on because the work process did not change. This was the case for George Metzger, who was still shaping steel rods in the blacksmith shop after World War II, despite having passed his eightieth

28. A 1940 aerial view of Tacony in Northeast Philadelphia. The factory is on the right and the town is in the middle, to the left of the trees. Today, much of the area is called Mayfair, but the original planned utopian community remains relatively untouched by the encroachments of modern society. (Tacony Public Library)

29. Map of Tacony in the 1890s. Tacony is bounded on the north by Cottman and on the south by Levick Street, two blocks south of Magee. In all, it covers about 500 acres. (Tacony Public Library)

30. An 1865 sketch of Tacony as railroad and boat terminus. Tacony came into being in the early 1850s with the building of the Buttermilk Tavern (at foot of flagpole) on the Delaware. Tacony was then a summer vacation spot for Philadelphians and local farmers. In 1872 it was connected to Philadelphia by railroad. Access by water and rail made Tacony attractive to Henry Disston. The Elm Tree Hotel was built to accommodate passengers. (*Disston "Bits"*)

31. Washington Tea House in Tacony, 1890. Henry Disston brought 200 steel-makers from Sheffield, England, between 1880 and 1900, which made English culture a central feature of Disston's isolated Tacony community. The tea house was owned by the Achuff family, shown in this picture. The building still stands at Hegerman and Unruh Streets in Tacony. (Atwater Kent Museum)

32. An early Tacony store on Torresdale Avenue, 1900. In 1903 a trolley line that ran north along Torresdale to Frankford was completed. Note the number of children in the car. Turn-of-the-century Tacony families were typically large, in part because of the better-than-average standard of living that a Disston job allowed. (Atwater Kent Museum)

33. Twin houses in Tacony. Henry Disston purchased 300 acres of land around the factory for his community. He divided the land into spacious lots and built more than 500 of these twin homes for workers to buy or rent. Workers could also purchase land to build their own homes. In this photograph, children on Tulip Street await the soldiers' parade honoring the men going off to war in 1917. (Tacony Public Library)

34. Tacony Park with railroad station in background, 1920. The park was another of Disston's provisions for the community, designed to enhance the quality of life for workers and their families. (*Disston "Bits"*)

35. Parade celebrating Tacony Week, 1906. The parade is on Disston Street below Keystone Street. Magistrate Thomas South's house is in the top left-hand side of the picture. (Atwater Kent Museum)

36. Children taking part in organized play outside the Disston Recreation Center, 1916. Discipline and order in children were necessary behavior patterns to instill in future factory workers. At this time, many boys between the ages of fourteen and sixteen worked in the factory, but in 1920 a law requiring school attendance until the age of sixteen was enacted. (Tacony Public Library)

37. The newly opened Disston cafeteria, 1919. According to one worker, the cafeteria was the "finest thing" the Disston company ever did for its workers. "What a relief it is to realize that I do not have to plan for that lunch every day, and arise a half hour earlier every morning to prepare it." (*Disston "Bits"*)

38. Men from the factory playing baseball during their lunch break in 1916. Employees organized teams that competed against each other within the factory and against other industrial teams. (*Disston "Bits"*)

39. Women from the office staff playing dodgeball during lunch in 1916. A number of athletic opportunities were open to girls and young women in Tacony. Disston fielded softball and track teams for women in 1919. Once a woman married, however, she was expected to quit work and focus on the family. (*Disston "Bits"*)

40. An 1890s Tacony baseball team in Atlantic City for a game. Tacony typically supported its team by sending fans to the game. Sports had an important social function in industrial towns like Tacony. (*Disston "Bits"*)

41. Tacony soccer team of 1898. With the help of Jonathan Marsden, English coach Sam Needham recruited many of these players from England. To get a job at Disston, it helped to be a sportsman as well as a smither, smelter, or file-maker. One of the Disston team's early opponents was the University of Pennsylvania. (*Disston "Bits"*)

42. Minstrel show in the Disston cafeteria building in 1920. These shows were popular events in Tacony as late as the 1950s, when they were held at the local high school. (*Disston "Bits"*)

43. Tacony's 27th Police Athletic basketball team, Disston Recreation Center Senior Champions in 1951. This team includes future college basketball stars (standing, left to right): Joe Gilson, part of the 1954 NCAA Champion La Salle College team; Dan Fleming, member of the Temple University final four team of 1958; and Harry Silcox, captain at Temple University in 1955. Coach and organizer of the team was Gregory Mead, center front row. Sports helped many a Tacony boy get a college education in the 1950s. (Atwater Kent Museum)

44. Tacony's 110 Field Hospital Regiment going off to war in 1917. Tacony's contributions of able-bodied men and Disston steel to the war effort brought company and community together in a common effort. (*Disston "Bits"*)

45. A 1918 parade of factory workers during their lunch break, in support of World War I. (*Disston "Bits"*)

46. Tacony residents turn out to welcome returning soldiers in 1919. The emotions of the community burst forth at the end of the war when the troops returned home. (*Disston "Bits"*)

47. The 1941 dedication of the Armor Plate Building. Disston's armor plate was important to the U.S. effort during World War II. In this photo, a tank made with Disston armor plate steel is on display. (Atwater Kent Museum)

48. William D. Disston at the Oxley Post American Legion, 1942. Here two World War I cannons are donated to Henry Disston & Sons as part of a World War II scrap-metal drive. (Private Collection of William L. Disston)

49. A 1941 *Life* magazine cover showing Disston workers waving flags at the Armor Plate Building ceremonial. This picture was one of the most famous of World War II. The government used it in magazines, movie theaters, and newspapers to show that American workers supported U.S. entry into the war. (Atwater Kent Museum)

50. Disston chain saw in a parade during World War II. Because Disston was the only American manufacturer of portable two-man chain saws at the time, it was granted an exclusive contract for a product crucial to the U.S. war effort, especially in the overgrown tropics of the South Pacific. Here Al McCloskey and Billy Reilly demonstrate the saw's effectiveness. (Atwater Kent Museum)

51. World War II parade in Tacony. Boy Scouts and the Disston chain-saw truck march down Knorr Street next to the public library. Many local communities in Philadelphia sponsored such parades during the war. (Private Collection of William L. Disston)

52. Workers and guests parade through the factory in 1942 on their way to a ceremony in the Armor Plate Building at which Henry Disston & Sons received the Army-Navy E pennant for wartime service. (Atwater Kent Museum)

53. Speakers' platform and workers at the Army-Navy pennant ceremonies. (Atwater Kent Museum)

54. Disston's oldest employees are given a front-row position during Army-Navy pennant ceremonies. The men pictured here each had more than forty years' working experience with the Disston company. (Atwater Kent Museum)

55. S. Horace Disston presents George Metzger with a gift at retirement ceremonies in the late 1950s. Metzger worked at Disston for more than sixty years. William Rowen, master smither of saws at Disston, praised the Disston family for their loyalty to their workers. "At Disston there was always a job for a loyal employee, despite his age. If you gave the Disstons a good day's work, you could expect their support when you needed it." (Atwater Kent Museum)

birthday. Machinery could not speed up or change this process, so production of circular saws and saws in general changed little during the time Disston & Sons operated.[27]

These procedures and opportunities gave the factory worker of the 1920s to the 1950s a chance for a better life. With each promotion, workers could increase their standard of living, purchase a house, and support a family. Factory wages were sufficient in a world that had few cars, few electronic devices, and few environmental concerns.

When one compares the life of the Disston worker with the textile workers described by Philip Scranton, there are similarities. Among both groups, Philadelphia workers were more productive than their national counterparts. Both focused on quality goods. Disston workers focused on high-quality steel and saws, while the city's textile workers made high-value damasks and curtains, pushing the financial returns for the local cotton sector far ahead of those for staple curtains. For Philadelphia textile workers in every category, both earnings and accumulation of wealth were at or above the average for the five largest cities in America. Disston management also paid a high wage to skilled workers. The difference between saw and textile manufacturing was that Disston was able to capture and monopolize the industry, while the textile industry of Philadelphia faced severe competition. Henry Disston & Sons' business was based on a consistent need for saws from around the world and a virtual monopoly of the saw-making industry. The textile industry's business was based on urban demand for seasonal and specific fabrics. This resulted in peak periods and layoffs among textile workers and steady employment for Disston workers, enhancing Disston's ability to plan wage scales and budget for the factory's needs.[28]

In the nineteenth century, production increases in the saw industry came from the invention of machinery to mechanize certain—but not all—production processes. Handles could be made faster with a power bandsaw, but no machine could speed up the tensioning of circular saws. Once few production improvements remained to be discovered, attention turned to eliminating the waste caused by the lack of communication between shops and, above all, turnover, training problems, and absenteeism.

The ongoing labor turnover crisis in the factory focused on labor shortages in certain job areas. Disston's twentieth-century Personnel Department cited two reasons for the shortage: workers being fired for poor performance, and workers quitting because they did not like the job. This simplistic analysis fits the primitive methods used in the early personnel offices then popping up all over the nation. The only evaluation done during hiring consisted of sizing up a prospective employee "as to looks," after which someone took the newly hired worker to see the job and await results. The majority of people who tried a job for a short time turned it down. This prompted the Personnel Department to report that it was "hard to tell whether a man is going to accept a position when he goes in to try it." It observed that since most of the workers were beginning in the least desirable jobs, Disston had to expect some job dissatisfaction and quick turnover. Harry Gorman, in the Heat Treatment Department, pointed out the severity of the problem, observing that "nearly all the men brought into the hardening shop to do sawdust work leave." Gorman recommended that the Personnel Department reinterview new workers after two weeks on the job so dissatisfied people might be redirected to work in another shop. "If a man is not followed up he thinks the quickest and easiest way out is to quit."[29]

Because of company practices that held on to skilled labor during hard economic times and promoted the able workers from the rank and file, there did not appear to be a scarcity of labor for skilled jobs. Unskilled, dusty, and dirty jobs, however, went wanting. Work was so uncomfortable in the handsaw sanding and splitting shop that only boys just out of school could be found to work there. Described by Albert Walton of the Costs Department as the "kindergarten" of the factory, the finishing shop lost nine of thirty workers in three months in the 1920s, but that "was to be expected because most were sent there to see how they made out." Success and good work by young laborers allowed them to transfer to more responsible jobs in other shops at higher pay. Hence the great job turnover in the shops that had "green labor and school aged boys."[30]

A newly enacted law requiring school attendance until the age of sixteen cut into the youth labor market in 1920. Before that time, the factory had hired a large number of schoolboys between the ages of fourteen and sixteen. Many of the fourteen-year-olds worked in the office as runners, or in the grinding shop pushing saw blades into a return cycle into grinding machines. Their salaries were $4.50 a week, with two days paid vacation and a chance for advancement if they "worked out." "Working out," according to one of their supervisors, meant "arriving on time, following directions and showing some interest in learning their job."[31]

Women and girls were most commonly found shellacking handles in the saw department. In 1920 seven of eighteen members of the Hack Saw Department, mostly young women, were transferred out because there was no work. They were sent en masse to the file shop, where they filled in as handle-finishers, shellacking the handles before placement on the file.[32]

The general report by the plant's Labor Department in 1919 indicated that there was ample help at the factory to meet current orders, except for skilled mechanics and young boys. Yet management felt these figures were understated because many departments hesitated to ask for help. William Disston was concerned that an entire department's production could be held up for want of a five-hour worker. To him, this was false economy; maximum production required enough labor that production lines would not be shut down. In this imprecise, best-guess situation for management, the labor supply remained Disston's primary unsettled issue.[33]

One way to increase the specialized skilled labor force was to enlist boys with good work records in training for one hour of each workday, a practice then being used in other factories. Much like the apprenticeship program, this procedure would draw from those already hired. The difference was that the training could start earlier and be done in specific skills chosen by management. This was the best insurance against labor shortages. The company put the plan into action by creating inducements for boys to enter a trade, and higher skills were taught to the boys in each department.[34]

Harry Dorsey of the Personnel Department advocated another approach. Many workers quit Disston because of sickness, and many women left their job for marriage:

> In a good many cases where the turnover is small there are mostly skilled labor and men who have learned their trades in that Department. They do not often want to quit. In departments where turnover figures are particularly high there is usually unskilled labor, who will stay there as long as they cannot get more [money] at another place.[35]

Nevertheless, a few skilled workers left Disston. The handsaw machine filing shop lost seven of twenty workers in 1920. One of the seven was a skilled man who left to open a machine shop with his father. The Personnel Department visited his home, something not done for laborers, in order to try to get him back, but to no avail. Two other workers were just being broken in when they left for higher pay at a shipyard. At the same time, the wood frame shop lost nineteen of forty-five skilled planers, bandsawyers, jimpers, and hewers to the Shipe Victor Talking Machine Company. Most left for higher wages. Such wholesale raids on skilled labor forced William D. Disston to evaluate the company's pay scale. He considered it important to pay higher wages for skilled workers because in the long run this would secure and guarantee their tenure.[36]

The increased competition for skilled labor in the Tacony area was confirmed by statistics compiled in 1920 by Harry Dorsey. Dorsey recommended that the company consider an increase in pay for all Disston workers so the company could recruit sufficient skilled and unskilled workers. William D. Disston replied "that the matter had been very carefully considered by the family and that it had been decided to deal with the men by departments." George Gebhart, chairman of the Factory Committee, also played down the importance of Dorsey's report, stating: "The fact that we had increased our hands by nearly 100 since January 1st showed we are meeting the situation." Besides, awarding pay raises by departments was less expensive to the company, since it currently paid higher wages only to workers in the most labor-competitive areas. Issues such as the general laborer's low salary were never considered by Disston management simply because the company could always find another warm body.[37]

Another suggestion for improving labor stability in the plant was to have the foreman in each department work closely with each newly hired worker. As the following report attests, this did not work in practice:

> A foreman, many times, while he should pay a great deal of attention to his men, is apt to overlook them and a Labor Dept. properly organized will have a number of men circulating around and among the new employees to see that they are properly placed in their departments, and if they have any trouble, any in getting along with the other men or in not having the proper support of the foreman in instructing, should be referred to the Labor Dept. and straightened out.[38]

Yet Disston management never financially supported the development of such a personnel staff. The truth of the matter was that the schools of the time were structured to provide a new and fresh supply of "laborers" or non–high school attenders each year. There would always be children who did not like school and could be encouraged because of family finances to accept work at the factory. Before 1945, schools were content to allow many students to drop out, making high school graduation an achievement for the few.[39]

The lack of support for the Personnel Department allowed primitive and unproductive personnel practices to prevail at Disston & Sons:

> Sometimes they have 12 men at one time in the room and each hears the other man being interviewed for a position. One morning, for instance, there were 5 boys interviewed for a position. They all came together and were apparently friends. Mr. Moore

> told one boy what the position paid and explained the job to him, but the boy refused the job. Mr. Moore interviewed a second boy, but he—probably due to hearing the other boy—also refused, likewise the remaining three all refused. If each one were interviewed individually and had not heard the other boy turn the job down, they probably would have been satisfied with it, but they saw the first boy turn it down and therefore would not accept the job.[40]

Dorsey recommended separate interviewing to give interviewers a chance to explain Disston's special benefits—the cafeteria, rest room facilities for men and women, the Keystone Beneficial Association, plant social events, and safety programs. The company could also use these rooms to conduct exit interviews. William D. Disston agreed, stating that this has been "the best report we have ever heard" and requesting that all personnel information discussed be forwarded to the general shop foremen for their reaction. What happened to the report is uncertain, but interviewing rooms were not constructed in the personnel department.[41]

After the labor supply, the issue most discussed by the factory committee was the quality of the steel. A view of Disston's advertising practices explains why. Disston & Sons were innovators in marketing and advertising in the nineteenth century. Their use of block prints, widespread publication, and the opening of regional sales offices throughout the nation helped to establish the firm as the nation's premier saw manufacturer.[42]

By the 1920s this advertisement campaign had been institutionalized into a standard message to the public. Disston saws were "Quality Saws," the standard of the industry, "the saws your grandfather used"; Disston made "a saw for every need."[43]

In the nineteenth century, sales offices could be found in eight cities, each acting as a regional center for complaints and personal sales meetings with customers. After World War I, this organizational pattern changed. A map in the October 20, 1920, *Philadelphia Made Hardware* catalog (published cooperatively by Miller Padlocks, Enterprise Industries, Plumb Company, Yankee Tools Inc., and Disston Saw) indicates that Disston had an active advertising campaign in every state in the union (see Fig. 24). The basic media used were magazines such as the *Saturday Evening Post, Literary Digest, Popular Science Monthly, Scientific American, Popular Mechanics,* and twenty-nine leading farm publications, with a total circulation of approximately 8 million copies each issue. Such a mass media plan negated the need for regional offices, except in lumber areas like Toronto, Seattle, and Guilford, Australia.

Because high shipping costs required Disston to locate repair shops closer to the sawmills, these three cities enlarged their sales operations to include repairs, and by the 1930s they had become small production plants.[44]

The regional offices not in the sawmill areas were systematically closed by the company. In their place a new sales strategy emerged. The hardware stores of America would become Disston sales agents by using Disston saws exclusively. The message to the hardware distributors of the nation, through the medium of *Philadelphia Made Hardware,* was that it was best for them to "concentrate on leading brands." An article by Smith Brothers, of Fayetteville, Arkansas, tells how they "disposed of duplicate brands" of saws and specialized in the Disston line because it was more profitable:

> We find it far more profitable to carry only one brand of saws than it would be to keep a lot of duplicate brands in stock at all times. . . . When a customer comes and asks for a saw, we center our sales ability on selling him a Disston saw. . . . I have found that most mechanics are partial to a Disston saw. It is a good saw, it is nationally advertised and there is a saw for every purpose. Many men who come into our store do not ask for a saw; they specify a Disston Saw. This was another thing that led us to "weed out" our other lines of saws and to specialize on Disston.[45]

The Smiths went on to show how carrying one saw line cut down on insurance. The store inventory was always less when a storeowner carried fewer brands, and less stock meant less insurance. Smith warned that those who had one line of products must select the "quality" goods, because cheap products resulted in dissatisfied customers.[46]

Disston fostered this approach, which would give the company a virtual monopoly over saw sales. And because this entire sales strategy was based on the quality of steel in Disston's saws, saws had to continue to be superior.[47] No one knew this better than the Disston management. The problem was how to convince the worker. This became clear to Disston's Factory Committee as members continually encountered weak and defective steel at the production level. The usual cause was slow shipment of raw materials to the steel plant, which forced the use of inferior steel that should have been discarded. This was the case in September 1919, when weekly reports indicated that the supply of good-quality steel was not keeping up with the production requirements throughout the plant. Under these restrictive circumstances, each department had to cut back on requests for finished steel.

The saw department was the first to feel the pinch. Some of the finished steel being produced during shortages was so bad "that a large percentage was returned as defective material," Dick Charlton of the Materials Committee lamented. "Most departments were running behind their production quotas."[48] The same year, the plant was also behind on deliveries of ship saws—specialized, sharp-pointed tools. They were difficult to manufacture because their fine points required special hardening. The company had 600 dozen such saws awaiting grinding, but it needed 400 dozen more to fill current factory orders.

William D. Disston was concerned about the possible reasons for the production lag. Were the delayed orders the result of poor planning? Elmer Roberts, foreman of ship saws, responded to Disston's query by pointing out that the increase in ship saw orders was unexpected and "due to the growing oil industry, as practically 90% of their orders call for ship saws." And ship saws required the firm's best grinders and mechanics. "Such skilled men were not always available, since they were often working in another part of the plant." What better evidence could there be of a continued shortage of skilled workers? Moreover, the narrow point of the ship saw needed extra temper so it would not break in use. "Our experience shows," concluded John Quinn, foreman of the sharpening shop, "that we have more trouble with ship saws than any other kind." William D. Disston directed the Engineering Department to develop machines that would allow for quicker setting of ship saws. Two months later Arthur N. Blum of the Equipment Department reported that new machines were in place and functioning.[49]

Poor steel also caused conflict between shops. The hardening shop claimed it was the steel works that weakened the steel; the steel workers put the blame on the hardening shop. William Disston ordered an investigation of the matter, with a report due back to the committee at the next meeting. On October 13, 1919, the findings were announced:

> [The claims made by] the hardening department have been substantiated and . . . the steel works have been discredited. Tests made on sheets furnished by the steel works that were ground before hardening show that they will harden clean and temper flat thus proving that the trouble is due to the decarbonized surface of the steel. It is therefore evident that the Steel Works [which did the decarbonizing] . . . is at fault.[50]

Disston ordered the Steel Works to find out what was going wrong and correct it. Harry Gorman reported on November 17 and 26, 1919, that the steel was showing a better quality, that conditions in the hardening shop were good, and that a large production yield could be anticipated.

Assurances were made that the issue of proper hardening would continue to be a major concern of the department.[51]

Nevertheless, poor quality persisted as a concern of the Factory Committee. In March 1920 John Quinn reported that the use of files of poor quality and the wrong size in sharpening saws was causing poor workmanship in the saw-sharpening shop. Three members of the committee supported Quinn, and the matter was referred to Dick Charlton for investigation. It was suggested that Charlton initiate tests of the files by the product-testing department. On May 25, assurances were given that the files were being tested and that progress was being made in getting files of the right quality to the saw-sharpening shop.[52]

In relation to this, William Disston spoke about the need for increased production and the responsibility of every worker to support the company in its quest for excellence. He submitted a plan "whereby our employees learn the work in several different departments" in order to conserve labor and maximize production. He asked for volunteers willing to work in other departments. Unfortunately for the company, few workers volunteered. In March 1920 William H. Batty let it be known that the saw and file shops still needed boys and laborers.[53]

Another complaint related to the quality of the saws was that inferior sandstones were being used to finish the steel in the grinding shop. During World War I, Disston & Sons purchased green sandstones, the only ones available. Because it was company policy always to have one thousand sandstones on hand, there were still some green stones at the factory. It was decided to continue their use when Charlton reported in October 1919 that the situation would correct itself in two months when the inferior stones were used up. This incident brings into question the Disston management's boasted concern for quality. Again and again the issue of quality was raised, yet decisions continually focused on cost containment.[54]

Absenteeism also influenced quality and production. William Disston was working on a plan aimed at a more thorough monitoring of workers who had excessive absentee records, and he invited suggestions to encourage better attendance. Disston told the committee that charts shown on a screen at the next "super" meeting in the cafeteria would depict the rise and fall of absenteeism in the company. Locally, Henry Disston School also focused attention on absenteeism, the *Disston Messenger* featured perfect-attendance students on its cover, and special lectures at the Tacony Library urged workers not to be slackers.[55]

Safety education was a more direct way to cut back on absences. Suggestions regarding plant safety were sought from Disston workers, and five finalists with particularly original and useful suggestions were awarded prizes and recognized in the "*Bits.*" In April 1920 a worker wrote of the danger to workers moving the large bandsaws on small trucks—the bandsaws slipping from the cart could spring loose, injuring the movers. The writer suggested that the Engineering Department look into the idea of building larger carts. Gebhart promised immediate action.[56]

The Factory Committee also discussed factory maintenance and the upkeep of the physical plant and machinery. William Disston recommended wharf repairs because of the deterioration of piers on the Delaware River side of the plant. During the next three meetings, the committee assigned the task of inspecting the piers and recommending a solution and reported progress in getting the work done. Over a six-month period, the wharf was repaired.[57]

Consideration of antiquated or obsolete machinery collected in various areas of the plant was ongoing. The first task was to inventory and remove it, the second was to find a locked storage area for items that might be useful in the future. The Machinery Committee was to undertake the task, keeping records of storage areas and usable parts.[58]

William Batty reported another important development: foreman training. Under the auspices of the Philadelphia Association for the Discussion of Employment Problems, a meeting was held on Monday, March 1, 1920, at the Stetson Auditorium at Fourth Street and Montgomery Avenue. About one thousand foremen were present, thirty-seven from Disston. The purpose of the event was to give "the foremen direct information concerning the vital topics which concern them as foremen, thus enabling them to get the viewpoint of other foremen and to see how problems which face them in their daily work are being solved by men in other industries." A lecture, "The Foreman's Job," was given by Charles Woodward of the Hydraulic Pressed Steel Company of Cleveland, Ohio. Batty described it as a "straight from the shoulder" description of the foreman's job in the steel industry. E. J. Cattell, city statistician, gave a speech entitled "If I Were a Foreman," with statistics on Philadelphia industries. William Batty reported that the Disston foremen were pleased with the program and looked forward to the next meeting. The meeting itself was felt to be an encouraging recognition of the importance of the foreman in factory management.[59]

The Executive Committee represented a sophisticated and forward-looking approach to managing a factory in the 1920s. Students of industrial management find few examples of companies promoting quality through decision-making by the workers nearest the production lines. Most argue that companies in that period practiced "Taylorism," relying on experts to decide the right thing to do. At Disston, however, statistics were used to reveal production weaknesses, and discussions were held with foremen to decide a course of action. Those closest to the product had the right to reject poor-quality work. Could it be that within the ranks of successful factories using skilled labor, decisions about work to be done were often initiated by the workers? Had the many years of paternalistic control in Tacony forged a bond between worker and owner that permitted such interchanges? While the social dynamics of the situation remain something of a mystery, Disston's management practices raise a number of questions about present interpretations of the history of industrial management.[60]

William D. Disston's cooperative management system increased the worker's respect for the Disston family and simultaneously gave the owners a true picture of production problems at the plant. This problem-solving group combined the insights of workers and management. William Disston was considered by workers to be one of the finest bosses in Philadelphia. He was respected for his knowledge of the plant, his willingness to listen, the soundness of his judgment, and his interest in the workers. Once a week he would tour the shops, stopping to encourage workers and shake hands with the foremen. For a young William Rowen he was "easy to talk with and down to earth." John Southwell recalled, "He knew the saw business inside out, and the workers knew it." Unlike those who followed him, William Disston devoted himself to the everyday routine of running the plant. As we shall see in the next chapter, his successor would be handicapped by a lack of such hands-on management.[61]

The son of Samuel Disston, S. Horace Disston, was financial manager of the firm in the 1920s. He did not share William's congenial reputation and could be abrasive and brutally direct. A worker who spoke with him had better know what he was talking about, for Horace would question and probe his every statement. These characteristics would prompt the family to vote him president and chief operating officer in 1939.[62]

It was S. Horace Disston who controlled the size of the workforce through the 1950s. Ignoring recommendations from the Factory Committee, he sidestepped the labor shortage by continually reducing the number of workers after World War I. As workers retired they were not replaced. More operations in the factory were being mechanized, and cutbacks in the workforce increased profits. Yet mechanization also meant more skilled labor, and such workers were both difficult to find and required factory training.

Nevertheless, Henry Disston & Sons was in its heyday. A company publication sent to the lumbermen of the country in 1923 claimed that Disston was the premiere saw producer in the world. The Disston saw works, it boasted, had built the first automatic machines for toothing saws. It had devised and invented techniques to harden saws under specially designed dies, thus keeping the saw flat; it had discovered how to temper saws under hot dies, thus ensuring uniformity of temper; it had made the first inserted-tooth circular saws for sawing metal, and it had also introduced into the United States band-sawing machinery for cutting saw handles.[63]

The Factory Committee lasted throughout the decade of the 1920s and ended with the coming of the depression. For a large firm in that period, it was a progressive form of factory management. William Disston's wish to work cooperatively with the factory workers fitted the Disston philosophy without interfering with the company's need to make a profit.

Disston & Sons' reputation as a company with a humane management system was popularized in the press and in the community in the 1920s. The company philosophy was that "a man will be happier in his work if he knows that he is a real part of the concern and if he knows that he can spend all his working days with you without danger of an overnight discharge." Management strategy included sports teams, newspapers, social events, and beneficent societies, but not a union. Large families, small-town amenities, and large numbers of skilled laborers further discouraged unionization.[64]

The leadership of William D. Disston raises questions about the training of Disston men for their role as leaders in the company. For the family there was only one way to prepare to become an executive in the company, and that was the way Henry Disston trained his sons: a three-year apprenticeship with the skilled craftsmen of the firm. Each Disston man started as a trainee under a department foreman, moving from foreman to foreman until he understood every process in making saws. The training lasted from two to three years, at which time the young Disston got a suitable management position and a party was given for all of the workers and foremen who had helped the Disston boy become a man.

The system had a number of advantages. The workers took pride in their training and knew the bosses well. It helped to develop emotional ties between the workers and family members and provided Disston executives with an understanding of how the factory worked. The Disstons also learned who was a good worker and who were the laggards. However, this system also encouraged a provincial leadership that was bound by tradition to the processes they understood and to a single vision of the company as a saw business. There was little planning for an emergency in which a cash-flow shortage would coincide with a need for new equipment. Lack of financial training and an understanding of how other industries and outside forces would impinge on the saw business also hindered Disston's leadership. Eventually, when Jacob Disston Jr., a stock market specialist and relative outsider, became president of the company in 1950, this leadership training program split the family into two camps—those who had been trained at the factory and reflected the traditional Disston understanding of factory practices steeped in quality and workmanship as learned through apprenticeships, and Jacob's followers, who believed that financial considerations mattered most in the business. Nevertheless, this training program lasted until 1940, when William Leeds Disston completed his training at the plant and moved into management.[65]

The Disstons balanced their interest in the plant with an equally active interest in developing the northeast community. As powerful members of the Tacony Manufacturers' Association, they were always ready to play a role in any scheme to enhance and expand that section of the city. The family still owned large tracts of land, and increased population meant increased value.

In December 1921 the Tacony Manufacturers' Association met to discuss the expansion of the Northeast. They proposed locating the Sesquicentennial Exposition of 1926 (then being planned) on the Roosevelt Boulevard, above Cottman Street, on a tract encompassing part of the Pennypack Creek. This 1,000-acre tract was mostly farmland, 50 percent of which was owned by the city and thus could be acquired at a low price. The association claimed the site had two advantages over every other site in the city: level ground for the economical construction of the large main buildings, and trees, water, and natural beauty for amusement buildings, sports, and attractive walks without any undesirable settlements or districts nearby. Transportation to the site presented no difficulty, since the Pennsylvania Railroad was adjacent to the area, and an extension of the elevated beyond the planned terminus at Frankford Avenue and Bridge Street would open the area to high-speed transit via a surface line up Bustleton Avenue from Bridge Street. Provisions for housing could be made in the open areas around the site. The Delaware River was also available for yacht and hydro-aeroplane transportation, and this anchorage was far superior to that of the southern portion of the river, where the dirty water discouraged long stays.

The instigators of this plan all had something to gain from its success. The president of the Tacony Manufacturers' Association was E. A. Gillinder of the Gillinder Glass Company; the vice presidents were Kern Dodge (Dodge Steel), and F. W. Daniel (Quaker City Rubber); and the list of directors included Edmond B. Roberts (a member of the Disston family by marriage), John F. Krauss (L. H. Gilmer Company), and C. C.

Fitler (E. H. Fitler Company), all representing northeastern firms located along the river. Their vision of progress included developing northeastern farmland that lay away from the river, which would give them a greater labor pool and keep down manufacturing costs.[66]

Eventually the city chose South Philadelphia for the site. The warship and marine shipping exhibits required the use of the existing Navy Yard as a base, making the selected site more naturally advantageous. Moreover, the Northeast was just too far from Center City to be seriously considered.

An important issue of the 1920s was the opening of a ferry between Tacony and New Jersey. After years of planning and negotiation with New Jersey officials, a Tacony-Palmyra ferry was launched on May 6, 1922. Service was delayed when the *South Jacksonville* (which later became the *Tacony*) was caught on a sandbar in the Delaware Bay. But "Bad begun was well done," as a timely adage said.[67] Initially the ferry was intended to serve two purposes: provide access to the White Horse Pike for cars going to the New Jersey shore from the northeast section of Philadelphia, and give the farmers in Burlington and Camden counties access to areas such as Frankford, Tacony, and Germantown for their produce. The ferry became important to the development of the northeast section of the city because it forced the city to improve the roads of the area. As late as 1925, Frankford Avenue was paved only from Bridge Street to Longshore Street, and Torresdale Avenue had just been paved to Cottman. Once the ferry was started, Cottman Avenue was paved and became a major connection to the western sections of the city.

There were two ferry boats, the *Tacony* and the smaller *Palmyra*. The *Tacony* could carry 36 cars on its lower deck and 600 passengers on a top deck, while the *Palmyra* carried only 20 cars and 50 passengers. An indication of the use of the ferry can be found in statistics for the first Sunday in April 1924, when the Tacony-Palmyra ferries carried 2,000 cars and 3,000 passengers. During that year the ferry carried more than 100,000 cars. Demand for a bridge was inevitable. Headed by Charles A. Wright, a company was formed to build a bridge across the river at the ferry site. On March 27, 1928, engineer Ralph Modjeski began construction on what became a $5 million undertaking. Two years later, the bridge was opened to much fanfare, and a permanent connection was formed between the communities of Tacony and Palmyra.[68]

The significance of the Tacony-Palmyra Bridge was that it, along with the Roosevelt Boulevard and the Market Street elevated train to Frankford, opened up the Northeast for settlement on a large scale. The transition was final and clear; no longer would Tacony be considered a community separate from Philadelphia. It was part of the city.

As for Henry Disston & Sons, Henry Disston, son of Hamilton, replaced an ailing Frank Disston as president in 1928. At the same time, William D. Disston became first vice president of the company. Within three years, Henry suffered a stroke and was unable to travel to Tacony from his Center City apartment, making William D. Disston the operating manager and CEO. The company's economic problems caused by the Great Depression would be the responsibility of these two men.[69]

CHAPTER

7

The Depression, New-Product Development, and World War II, 1929–1945

LIZABETH COHEN describes the decade after World War I as a period when programs based on industrial welfare were used to focus worker attention away from unionism and political activity. Employers felt that these programs both forestalled union activity and guaranteed worker loyalty. But over the next decade, as a result of a wide range of social and cultural experiences, the American working class underwent a gradual shift in attitudes and behavior. Industrial-welfare schemes eventually failed, and while the Great Depression rocked the nation such unifying social experiences as radio, movies, chain stores, and the fireside chats of President Franklin D. Roosevelt coalesced workers behind the Democratic party. "What mattered most in explaining why workers acted politically in the ways they did during the mid-thirties is the change in workers' own orientation," not what employers did or did not do. At the heart of the issue was the Great Depression, which propelled this new unified workforce into actions that resulted in the organization of Disston's first company union.[1]

The effects of the 1929 stock market crash were not felt at Disston until about 1931. Rapid decline in construction work reduced the market for lumber and thus for tools to some extent. When carpenters faced 50 percent unemployment and hardware stores closed by the dozens, Disston was bound to suffer. These circumstances led to a drop in the workforce from 2,500 in 1925 to 1,400 in 1933. Under these circumstances, pensions meant little to the laid-off worker. Industrial welfare's failure to convince workers that they had a long-term future with the company doomed pensions as methods of tying employees to the firm. For example, Disston's management considered its pension plan sufficient to safeguard workers, but this was true only if the relationship between the company and the employee was stable. This was not the case in most companies in the 1930s, including Henry Disston & Sons. Employers were looking for long-term commitment, but workers were looking for immediate wages.[2] The Great Depression caused employers to reconsider the promises of industrial welfare and caused workers to conclude that employers valued these programs only when they were convenient and cheap. Clearly, the lack of revenue in the 1930s turned the Disston management's attention away from industrial-welfare programs.[3]

Layoff patterns during the depression also encouraged unionization. At Henry Disston & Sons, unskilled middle-aged workers were likely to lose their jobs to younger workers, who could be paid less and still do the same jobs with minimal training, while the company made every effort not to lose skilled smithers or skilled steel-makers. This helped to make seniority rights a major union demand for the unskilled but a less important issue for the skilled. The workers kept on at the plant tended to be steel smelters, while those let go were likely to be laborers.[4]

The different interests of the laborers, on the one hand, and the steelworkers and skilled smithers, on the other, led to a long-term labor dispute that would not be settled until the union strike of 1940 changed how Taconyites lived in their community. The bitter fight between the unions representing these groups is discussed in Chapter 8.[5]

The lack of either a job or unemployment compensation left workers to deal with the depression on an individual basis. As orders fell off, laborers and nonskilled workers were let go, with skilled workers filling their jobs or put on part-time schedules, sometimes at reduced laborer salaries. Saw-smither William Rowen supplemented his part-time salary with whatever work he could find. He had no money for carfare or anything extra in the house. His entertainment on Saturday night was to walk with his wife to Frankford Avenue and Rawle Street and then back home. The new Liberty Movie House on Torresdale Avenue remained half empty on Saturday night because many people in the town could not afford the 10-cent admission charge. At Darreff's grocery store, a long list of residents who owed money was kept next to the cash register. Deliveries continued, and eventually almost all bills were paid. The lone exception was a women who ran a boardinghouse on the corner of Disston and Keystone Streets who never paid off a $2,000 debt.[6]

William Rowen, a skilled worker, believes to this day that Disston was one of the best companies to work for, even during the Great Depression.

He was told: "If you give the company a good day's work and were honest, once hired, you always had a job at Disston." During these difficult depression years, he saw the benevolence of the firm at work. Disston would advance wages so workers could pay debts; no Disston mortgage was foreclosed, and late rent payments were accepted without penalty during these years.[7]

Yet for a minority of Taconyites, Disston & Sons was a heartless organization run by a family that cared only about money. As money became tight at John Hansbury's house, he and his father, a plumber, did their own repairs, and Hansbury lost his house at 6606 Torresdale Avenue in 1931 because he could not afford the mortgage payment. He and his wife moved back in with their parents on Ditman Street for the duration of the depression. Hansbury worked at Sears at the Torresdale golf course and in the Civilian Conservation Corps. Finally, in 1938, he got a job as a holder in Disston's Circular Saw Department. He worked with a smither, turning the circular-saw blade as the smither circled depressions, lined high spots, and used different hammers to bring the saw tension into proper alignment. He stayed at Disston for only six months because "Disston paid only 50 cents an hour and the union was too weak to do anything about it."[8]

Hansbury's opinion was not shared by other unskilled laborers. George Gross, a worker in the hammer, armor plate, and smelting shops, found the working conditions at Disston & Sons "excellent." Disston took care of the workers, providing medical care and a cafeteria with inexpensive hot lunches. Gross cut his hand while fixing something at home and went to the local hospital for stitches. The next day at work he informed Dr. Valentine about his accident. Valentine, angry that Gross had not come to him for free stitches, made Gross come to him each day to have his wound bandaged until it healed. Gross felt he was treated fairly and enjoyed his work at Disston.[9]

For young people like William Hillerman, who graduated from Frankford High School in 1933, there were no jobs. Hillerman remembers the depression in Tacony as a time when young men between the ages of eighteen and twenty-five could be found every day at the Disston Recreation Center playing baseball, volleyball, soccer, or basketball to pass the time while they waited for jobs to open up. He eventually got his first job at Darreff's in 1939, and with the coming of World War II he got his first full-time job, working in Disston's new Armor Plate Building. In 1946 he was moved to the Chain Saw Department.[10]

These stories give some indication of the mixed sentiments in Tacony during these years of change and adjustment. For the Disston family, there were no dividends or profits after 1931, but the Great Depression did not kill the firm, even though the nearby Erben-Harding Woollen Mill and the Gillinder Glass Company closed their doors in bankruptcy. There were a number of reasons for its survival. First, its steel plant was making the best steel plate in the world, and orders continued to come in throughout the depression. Second, an increased emphasis on new-product development prepared Disston for the economic recovery of 1938–39, which was on a national level gearing up for World War II. Finally, William D. Disston's extraordinary administrative ability and the workers' belief in his honesty and integrity allowed the company the flexibility to adjust workforce hours without protests.[11]

New-product development had made significant progress in the decade before the Great Depression, when William D. Disston initiated a search for new products. In 1925 the Weyerhauser mill in Everett, Washington, needed saws that would cross-cut four-foot hemlock logs. Disston's reputation for specialty and custom work attracted Weyerhauser to order the saws from the Tacony plant. These orders continued, and in 1929 Disston hosted a luncheon for customer representatives, using as tabletops slightly smaller versions of the new saws.[12]

Maintaining its grip on the lumber-saw market worldwide, Disston now ventured into an untapped market. Disston's *Tool Manual for School Shops* (1927) claimed to be an authoritative text on the theory and use of tools. It offered school instructors and principals shop designs and a curriculum. Every tool pictured or discussed was, of course, made at Disston. The extension of shop programs across the nation at that time increased Disston's business.[13]

In 1930 a steel garden rake was put into production after a professional gardener presented a working model to Norman Bye in the engineering department. The Hunter Pressed Steel Company in Pottstown, Pennsylvania, assembled the rake, using Disston steel and Disston-bought handles. A lightweight hand pruner made of magnesium was developed at the same time and became an immediate success with home gardeners throughout the country.[14]

Sugar beet knives became a popular seller at Disston & Sons during the 1920s (see Chapter 5), when electric motors were used to create a machine that could cut and grind sugar beets into a pulp that was later used in sugar production. The knives for these machines were serrated at right angles every quarter of an inch to form a surface that could both cut and chop beets. Needless to say, these knives were extremely difficult to sharpen. The Disston management, a leading producer of these newly shaped knives, recognized an opportunity.[15]

Norman Bye, chief engineer, and Sam Freas, master machinist, were assigned to design a practical machine for sharpening sugar beet knives.

They did so in January 1933. Detailed explanations and drawings accompanied a set of applications to the law firm of Howson & Howson, specialists in patents and trademark copyrights, for the patent. Patent number 711,246 was given to the project, and lawyer John Howson was assigned to pursue the legal process leading to a final patent. A report was immediately returned to the Disston company that there were eight patents that might be impinged on by the beet-knife application. Of the twenty-five claims made by Freas and Bye, four were rejected as having been previously patented.[16]

The new sugar beet knife-sharpening machine consisted essentially of three parts: a stationary arbor on which the file was mounted, a knife holder that could be moved in front of the file arbor by means of a feeding mechanism, and a movable apron that raised and lowered the knife holder into the filing position. In correspondence between Bye, Freas, and Howson in the years between 1933 and 1939, the questions raised in the patent process were answered, and Disston was given the patent rights to the beet-knife sharpening machine process in America.[17]

Realizing the difficulties, when working by hand, of keeping the cutting edges of the sugar beet knives uniform and sharp, Freas and Bye eventually perfected two separate machines for doing the work automatically. These machines were designed to be sturdy, easily adjustable, and automatically operated. They enabled the operator to file the blades with uniform bevels. This prolonged the life of the knife, produced uniform cossetts, and afforded better extractions—all of which got more sugar from the beet.[18]

The beet knife blade and sharpening business was to become one of Disston's most profitable product lines in the 1940s, 1950s, and 1960s. One has only to read the list of machines sold in 1934 or the continuing orders from Belgium, Japan, England, and Sweden, as well as the American market, to realize the significance of this product to Disston's profitability. However, as in the case of the chain saw, the profits plummeted after World War II. In the 1950s a German machine took over the market because it was less expensive, smaller, and simpler to operate. Nevertheless, for decades this machine showed off the skill and craftsmanship of the workers in the machine shop, who not only made machine parts for Disston but also produced the sugar beet sharpening machines.[19]

Another important part of the story was the development and perfection of products. In the early twentieth century, factory inventors like Bye and Freas relied on trial and error, just as Alexander Graham Bell and Thomas Edison had done years before. Complaints about specific problems with the machine can be found scattered through the beet knife file kept by Bye. He personally traveled to sugar factories throughout the country to see, firsthand, the problems cited by customers. It was clear, he noted, that "conditions under which these machines operate in the sugar mills are quite different from those in the factory." Visiting the Brighton and Logmont mills near Denver, Colorado, Bye found that the knife clamp used in the routing operation was coming loose during the sharpening operation. He discovered that mounting the knife at its ends overcame the problem. A visit to the Holly Sugar Company in Tracy, California, settled a noise problem caused by excessive pressure in the sharpening process. The continued need for these field visits tells a great deal about the use of machines in American factories. They operated well only when they were fine-tuned by an expert who understood their operation. One could not sell a machine and expect just anyone to set it in place and begin running it.[20]

Depression-era experiments on military steel plate also proved important to the company and the workforce. In the mid-1930s William D. Disston was approached by the Caterpillar Tractor Company of Peoria, Illinois, to provide tank bodies for tractor frames. It seems that the government of Afghanistan had ordered nine tanks and three tank bodies from Caterpillar. An agreement called for Caterpillar to ship to Disston tractors that would be modified to fit an armored tank body complete with one 30-caliber machine gun and a main turret with a 37 mm cannon.

In 1935 the finished order was shipped to Karachi, India, and from there by train to Kabul, Afghanistan, where the government declared a holiday and paraded the tanks in the city square. A second order for four tanks followed from the Chinese government. These were finished in 1938 and delivered one year later after military security problems were worked out between the two countries.[21]

By 1935 these new products and Disston's expertise in making armored steel enabled the company to operate in the black for the first time since the Great Depression began. This was good news for the Disston family because it ended four years without dividends and with the reduced salaries for the executive offices of the firm.

The same year, Jacob S. Disston Jr., the last grandson of the founder, entered the company as sales manager. His previous experience was with a stock brokerage firm that had gone bankrupt in the depression. Despite limited experience in the steel or saw business, he would be expected to oversee a nationwide network of lumber and hardware sales. His younger brother, Horace C. Disston, was at the time sales manager of the steel division, giving that side of the family complete control of sales.[22]

Two years later William L. Disston, the first member of the fourth generation of the Disston family, began work as an apprentice in the steel melting shop. William learned the saw/steel business as his father and grandfather had learned it before him. The training period lasted three years, during which time William worked in all departments of the steel, saw, and file works.

In 1937 CEO William D. Disston suffered a heart attack that required a six-month leave from the firm. This left the company with its two top executives in ill health and prompted a management reorganization in 1938. S. Horace Disston became president, Richard Nalle became first vice president, and Jacob S. Disston was put in charge of sales. William D. Disston returned to work at the end of 1938 as a vice president responsible for purchasing, advertising, and the plant in Toronto, Canada.[23]

The company's attention in the 1920s and 1930s to upgrading the quality of face-hardened, light-gauge armor plate for navy gun shields, aircraft, and light tanks proved beneficial when tensions in Europe raised the threat of war. Robert Sibley, superintendent of Disston's Armor Plant Division, made weekly trips to the armed forces proving grounds near Baltimore, Maryland, with plates for testing. Monitored by the U.S. Army Ordinance Department, these tests established Disston's reputation for superior armor steel. By 1939 large orders were received for face-hardened armor plate fabricated to different shapes and different thicknesses—for Army tanks, armored scout cars, and bombers, or for Navy gun shields. Government contracts were important to Henry Disston & Sons throughout the depression years. Preparations for World War II kept workers employed in Tacony.[24]

In 1934 William D. Disston and Norman Bye came back from the Leipzig Fair in Germany enthusiastic about the Maryfield chain saw exhibited there. They brought with them a German named Arthur N. Blum. Blum was put in charge of a new Products Department headquartered in the cafeteria building, which had been closed because of the depression. Never a part of the Tacony community, Blum lived in a hotel at Thirty-Sixth and Chestnut Streets. He was appointed chain-saw developer, with Al McCloskey as drafting assistant and Billy Reilly as machinist. The operation was simple. Blum's ideas were drawn to scale by McCloskey and built by Reilly. Their first assignment was to develop an electric two-man chain saw. All other projects soon fell by the wayside, and the division became known as the Chain Saw Department.

Two years after Disston's original electric chain saw was developed, Blum and McCloskey visited a Pennsylvania Railroad site, where one of the supervisors suggested that an air motor might be applied to the chain saw so the pilings used in bridges might be cut under water. By 1938, production of air-powered chain saws had begun. The government military departments purchased Disston's electric and air-powered chain saws, but soon purchasers complained that large trucks were needed to carry the compressors or generators that powered the motors. They suggested that gasoline engines be used to power the saw, and they invited the Disston lab team to Fort Belvoir, Virginia, to meet with representatives from Kiekhaefer Corporation and the Corps of Engineers.[25]

Present at the meeting were Elmer Carl Kiekhaefer, president of a small outboard engine company in Cedarburg, Wisconsin, and Major C. Rodney Smith of the Army Corps of Engineers. Kiekhaefer and Smith had met in February 1941, when discussions were first broached with Kiekhaefer to develop a prototype of the German "Stihl" gasoline chain-saw engine. Major Smith showed Bye and Kiekhaefer the Army's only "Stihl" engine in the government's possession, and both parties agreed to cooperate in producing a similar, if not superior, American counterpart. Disston would manufacture the guide-rail and chain-saw shaft, and Kiekhaefer would produce the engine. This was a business partnership that greatly shaped future events at Henry Disston & Sons.

Within two weeks Kiekhaefer produced a set of blueprints for an engine the Army considered to be superior to the "Stihl." The engine was ready four weeks later and shipped to Fort Belvoir along with Disston's long-used guide-rail chain-saw system. The Disston-Kiekhaefer prototype was given to Major Smith, who arranged for a testing date for all bidders to the Army chain-saw contract.

At the test a second company, Reed-Prentice of Worcester, Massachusetts, presented a second chain-saw prototype, but their engine proved defective when fractures appeared in the crank shaft during the demonstration. The Disston-Kiekhaefer saw functioned perfectly, making them the preferred company for the contract. However, the report of the tests indicated that the Reed-Prentice saw mechanisms had advantages over the Disston prototype and that Kiekhaefer might consider a Reed-Prentice/Kiekhaefer team effort.

Henry Disston's size and reputation for meeting contract specifications, however, gave the Disston-Kiekhaefer alliance the edge. The War Department notified Disston that if a contract were offered on the Disston-Kiekhaefer bid, Disston would become the preferred saw-mechanism supplier and receive the entire defense contract, relegating the Kiekhaefer company to subcontract status. Learning of this commitment Carl Kiekhaefer was upset because his company, despite its small size, had shown itself to be superior during the engine development phase. The government's decision grated on Kiekhaefer enough for him to quietly enter a conspiracy with Reed-Prentice to remove Disston from contract consideration.[26]

In July 1941 Carl Kiekhaefer wrote his Washington, D.C., engineering representative, Jones Allan, about his feelings. "It is apparent that it is much easier to work with Reed-Prentice than with Disston," Elmer wrote. "We have also learned from our Philadelphia representatives that [Disston] is inquiring around for various engine parts, . . . apparently doing everything to keep from using our engine." All evidence indicates that this was untrue. In fact, Disston management was at that time enjoying the fruits of a governmental armor-plate contract and simply awaiting news of the Army chain-saw contract.[27]

A secret letter to F. W. McIntyre, vice president and general manager of the Reed-Prentice Corporation, was clearly anti-Disston. Kiekhaefer explained that Major Smith was being criticized by his superiors for not offering a contract for the power-saw project. Kiekhaefer reminded McIntyre that Smith's report on the bidder's test judged Reed-Prentice saw mechanisms superior to Disston's and that Disston had refused to make improvements on their long-standing prototype. "I shall," continued Kiekhaefer, "instruct our Washington representative, and I trust that it meets with your approval, to see Major Smith and to see what his reactions may be to a proposed inclusive Reed-Prentice-Mercury (Kiekhaefer) arrangement. We believe Major Smith will drop the Mercury-Disston arrangement." A visit to the Mercury Outboard Motor plant by McIntyre was arranged as Kiekhaefer stalled for time, keeping the option of a Disston-Kiekhaefer contract open.

Kiekhaefer's scheme of cutting out Disston from the contract was doomed from the start because of the size and reputation of Henry Disston & Sons and the near bankrupt status of Mercury Corporation, which was unable to produce outboard motors because of a lack of aluminum in wartime. Government red tape delayed any decision on the Army chain-saw contract until December 1941, when the attack on Pearl Harbor brought America into the war. The all-out war effort demanded that the Army have chain saws, and on December 23, 1941, Disston was given the contract; they immediately subcontracted to Kiekhaefer for engines.[28]

The Disston-Kiekhaefer chain saw was standard in the Army and throughout the nation during World War II. But after the war, other engine manufacturers entered the field, and as the technique of cutting with chain saws changed, Disston's high-power, two-man, heavyweight saw became obsolete. At this point Disston sponsored the design and tooling of one new large unit and one new small unit. The large unit was built by the Kiekhaefer Company, but the small unit remained in blueprint form. The delay was entirely the result of Kiekhaefer's lack of interest in the project. The fact that the U.S. government Disston chain-saw contract had saved his business during the war meant little to the eccentric and energetic capitalist Elmer Carl Kiekhaefer, who now was more interested in boat racing and Mercury outboard motors. Although Kiekhaefer's chain-saw engines did an acceptable job, they were not lightweight, and Disston had to buy them in lots of five hundred to obtain a favorable price. After a year of bulk contracting with Kiekhaefer, one room in the plant contained nothing but that firm's "Mercury" gasoline engines. Worker Russell McIntyre remembers men discussing the fact that two engines arrived at the plant for every one used in producing a chain saw. To the workers, it was clear the company was not making money producing two-man chain saws.[29]

Other engine manufacturers, in the meantime, had refined their units and revised them to keep abreast of the constantly changing requirements in the chain-saw field. However, Disston's contract with Kiekhaefer stipulated that Disston must purchase chain-saw engines in large bulk orders exclusively from them. This contract worked well during the war, when Disston had a virtual monopoly on chain saws but was unsuited for the new peacetime market. Disston management again appealed to Kiekhaefer to speed up development of a new lightweight engine and to allow smaller purchases of the larger engines so Disston might survive in the new chain-saw market. Kiekhaefer, with his usual skill, turned down the request for smaller orders and insisted that his company was moving as quickly as possible to complete the small-engine project. Disston's concerns remained unanswered.[30]

The drawback to Disston's chain saws, whether electric or air- or gasoline-driven, was that the power source was always too cumbersome to allow for ease of operation. The gasoline chain saw, for example, took two men to operate because of its weight and size. In the opinion of draftsman McCloskey, one reason for Disston's decline in the 1950s was the company's inability to develop a lightweight chain saw. Similarly, William L. Disston remembered his father William D. Disston lamenting that Blum's department moved too slowly in testing and developing any of their ideas.[31]

The contract with Kiekhaefer proved disastrous to Henry Disston & Sons. Disston's dependence on Kiekhaefer for engines allowed the company to remain captive to the whims of the supplier. This issue would force a chain of events in the 1950s that would have an impact on the financial soundness of the company.

Nevertheless, between 1939 and 1945 Disston was the only American manufacturer of a two-man portable chain saw. This gave the company the exclusive contract in World War II for a product crucial to the war effort, especially in the overgrown tropics of the South Pacific.[32] This was a remarkable achievement. The company pursued this and other new ideas even though orders were down and a negative cash-flow placed the firm in jeopardy. Credit must go to William D. Disston's leadership and unwillingness to lay off skilled workers, some of whom were moved to the development department "just to have something to do." The same men were so grateful for the work "that they cooperated more and worked harder for Bye."[33] The Great Depression, at least in Disston's case, stimulated new product development and innovation.

As early as 1938 the increased probability of war in Europe brought new orders for armor plate. In 1939 Disston spent $250,000 on a new power plant. In 1940 it offered to operate a newly leased building on its grounds for the government's purposes. City Council voted to allow the government to close off Disston Street between Wissinoming Avenue and the Delaware River for the plant site. The federal government paid the $65,000 necessary for the easement, and an armor plate factory was built. The first government order required the employment of sixty additional men. Tacony was being economically revitalized through war production. S. Horace Disston went so far as to claim, "Orders are coming in so fast that our current facilities cannot keep up with them." This surge, coupled with Midvale Steel Company's contract to make battleship armor was turning Philadelphia's depressed economy around fast.[34]

Concerned what perceptions this new armor plant might foster in the minds of hardware distributors, Horace Disston sent a form letter to them in July 1941. Stressing a stepped-up production schedule, Horace noted: "Our regular products continue to come in at a faster pace than we can manufacture." He then added, "The reason we cannot make deliveries on these items as promptly as you and we desire is not the fact that the manufacturer of armor plate is interfering with production." Nevertheless, it was clear that the country's gearing up for war was having an impact on domestic products.[35]

Apart from those rejuvenated by the war effort, other types of industries were moving into Tacony. Disston leased land to A. S. Beck Shoe Corporation for a $50,000 building. Income from this lease financed new machinery and equipment for conversion of the plant to wartime needs. The depression was over for Disston. The plant began a nine-year period of three-shift twenty-four-hour operation.[36]

Tacony continued to be a patriotic community, and its British flavor prompted sympathy for Great Britain. An August 1940 rally attracted one thousand Taconyites to the Disston Recreation Center in support of the British in their war against the Germans. The speakers advocated mobilization preparedness by the national government and increased military aid for England.[37]

During the same month, the Disston company was honored as a firm that fostered the highest type of industrial and employee relationship by the Forty-Fifth Annual Congress of American Industry. The company received special recognition for providing incentives for older as well as younger workers. Disston management paid the expenses of nine employees to attend the recognition ceremonies at the Waldorf-Astoria Hotel in New York City. Each had been with the company for more than forty years.[38]

Ground was broken for the armor-plate plant in October 1940, and it was built and equipped by the Quartermaster and Ordnance Departments of the U.S. Army, with members of the Disston firm acting as consulting engineers. The Disston company had been producing light armor plate for the government since the Civil War and was given the plant on a lease basis. This addition tripled the firm's capacity for fabricating and heat-treating armor plate. The process involved casting alloy steel ingots, blooming and rolling them into plates, cutting and machining these plates to fit the purpose for which the plate was intended, and finally heat-treating and straightening the plates to develop the bullet-resisting qualities essential to the finished product.[39] Disston's armor plate was used for many purposes: gun shields for all artillery pieces, from 37 mm guns to 155 mm howitzers; armor for scout and combat cars, for light and medium tanks, and for small naval craft; armor for pursuit and observation planes, bombers, and other military aircraft; and test plates for checking the quality of bullets and armor-piercing projectiles of various calibers.[40]

On June 4, 1941, William D. Disston was appointed chairman of yet another celebration. He and Norman C. Bye, George E. Hopf, and George E. Jeffrey Jr. were to plan a celebration around the cornerstone-laying at the new power plant and the dedication of the new Armor Plate Building on June 16, 1941. As part of the celebration, an M3 medium 30-ton tank from the Baldwin Locomotive Works in Eddystone, an M2 half-track car from the Autocar Company in Ardmore, and a 37 mm A.A. gun from the York Safe & Lock Company in York were put on display at Disston. All had in common armor plate provided by the Disston Steel Works. S. Horace Disston awarded 40-, 30-, and 20-year service pins to employees. Disston's grand old man, George Metzger, eighty-four-year-old foreman of the Blacksmith Shop, was honored for his 70 years of service. In all, 9 active employees received a 60-year pin; 54 received a 50-year pin; 124, a 40-year pin; 190, a 30-year pin; and

350, a 20-year pin—proof, said Disston, "that a man is entitled to employment so long as his ability and skill is unimpaired."[41]

S. Horace Disston made two speeches that day. At the laying of the power plant cornerstone he stated, "This plant marks an important milestone because its building has been a peacetime project. . . . The real purpose of this central power plant is and will be predominantly peaceful." At the armor plant, he said, "It is no news to anyone who works for Disston that we are busy with defense these days. . . . Nevertheless, I want to remind you that a most vital part of the defense job consists of *maintaining* a normal delivery of Disston saws and other tools to our regular customers, many of whom are busy in defense work."[42]

Pictures of Disston workers waving flags at this event were in *Life* magazine on July 7, 1941, and in newspapers sent around the world. In what became a classic wartime image, the picture of the flag-waving workers was placed inside the outline of a map of the United States. Such pictures were proof that America supported a new war effort and was gearing up to help England and France.[43] H. P. Aikman wrote S. Horace Disston that he had tacked the picture "on the wall over my desk." Another letter, from Joseph G. Terhorst, mill-supply buyer at the Terre Haute Heavy Hardware Company, commented on the opening of the Armor Plate Building in less than patriotic terms:

> We hope that this means that you will be able to take care of your jobbers in a much better manner than you have in the past. We realize that national defense comes first but we believe also that your old customers should be given a certain amount of consideration also. There are too many concerns that are thinking more about the large orders they can get rather than taking care of their customers who were loyal to them during the depression.[44]

Encouragement came from Edmund Orgill Brothers & Company, from Memphis, Tennessee. "So many factories have seemed to be holding back; but we are 100% for bringing this war to a successful conclusion as soon as possible and think it's just as much our war as it is anybody else's." Factories that have gone into the war "heart and soul" have done the right thing, he wrote. Hardware "doesn't make much difference as compared with this other problem." But W. F. Kennedy of Ott-Heiskell Company in Wheeling, West Virginia, warned Disston to keep "'first things first' and not indulge in so much talk and action about the maintenance of so-called social gain. . . . It is not a very complimentary tribute to our civilization when such a vast portion of our time and energy and substance must be employed in producing instruments of destruction. And yet we must be practical."[45]

Outside journalists remarked on the patriotism of Disston workers. Newspaper after newspaper carried the story: "In sharp contrast to recent pictures of striking defense workers waving demand-banners is this photo of flag waving workers, taken at the recent dedication of a new armor plate plant in Tacony, Pennsylvania. They are employees of Henry Disston and Sons, which expects the new plant to triple the firm's capacity for making armored plate for warships and other defense uses."[46]

In the factory, the war years brought a change in the types of products produced. Saw production dropped; military production increased. Most of the steel at the factory went to produce armor tanks, airplane seats, and rifles. Norman Bye found himself spending a good deal of his time in developing steel strength to make tanks and airplane seats, especially in fighter planes, less vulnerable to the impact of bullets and shells. Bye's work and that of the men in the factory led to special recognition for Henry Disston & Sons.[47]

The highest award a company in wartime could receive was the Army-Navy "E" Pennant, symbolic of maximum war production. In late October 1942, Disston was notified that the company would receive the award. A meeting was held on November 4 to organize yet another celebration. Hugh Green, president of the United Saw, File & Steel Products Union, spoke for labor, and S. Horace Disston represented the Disston company. Howard Pew of the Sun Oil Corporation was contacted to secure Lowell Thomas as master of ceremonies and radio singing star Margaret Speaks to sing "The Star Spangled Banner."[48] The committee placed an ad in the newspaper to announce the event. Worker William Massey suggested "picturing a mass of workers in background, and wording the story to bring out what the workers are doing." It was agreed that the event would be aimed at building up "the workers and not the company."

The ceremony was held as scheduled. Speakers included Acting Mayor Bernard Samuel and Hugh Green. Colonel D. N. Hauseman, chief of the Philadelphia Ordnance District, presented the pennant to S. Horace Disston, who commenced his remarks with "The 103-year-old-veteran has gone to war again."[49]

The war years brought to an end the Disston family's long-standing policy of renting homes to workers, which Henry Disston had first instituted. Mary had continued to rent homes and sell lots with the deed restrictions in place, keeping her husband's utopian experiment intact. Upon her death these properties passed into the Mary Disston Estate

Trust and all proceeds went to her heirs. The trust was to continue until all the family members named had died.

Catherine Seed worked for the estate for nine years (from 1936 to 1945), collecting rents and arranging repairs for some 365 homes in Tacony from the estate office at 4811 Unruh Street. These houses were mainly between Unruh and McGee Streets on Van Dyke, Hegerman, Tulip, Rawle, and Knorr Streets and the 6700 block of Marsden Street. They were rented only to Disston employees. As noted above, the Mary Seven, homes on a block of Wissinoming Street, were rented exclusively by black families. The rent ranged from $18 to $25 and was paid monthly at the estate office. When rent was not paid, a shop list was made up and paymaster Burt Castor would withhold pay weekly. Homes were repaired quickly by the company at the renter's request.

After the death of Mary Disston's last named heir in 1942, the estate was liquidated. By 1944 all homes had been sold. The trust adhered to the requirements that the house be offered to the lessor first and that it be put into good repair, including roof, plumbing, and even wallpaper. The price of buildings was set at $2,500 for the homes on Van Dyke Street, $2,700 for those on Hegerman Street, $1,875 for the ones on the 6700 block of Marsden Street, and $1,500–$1,600 for Rawle Street. The homes were quickly purchased by renters exercising their option to buy. Ultimately only one home was made available for public sale.[50]

The selling of the Mary Disston Estate was not the only sign that Tacony was becoming more independent of Henry Disston & Sons. A newly arrived Holmesburg rival was John J. Nesbitt's, a company that sold heating, ventilators, and air conditioners to schools. In the 1950s it would become the leading employer of Tacony youth.[51]

As for the town, the rapid industrial development and private-home building starts of the 1920s came to a halt with the Great Depression. This meant that there was little change in the town's boundaries between 1932 and 1950. The highly populated area was east of Cottage Street stretching to Keystone Street between McGee Street and Princeton Avenue. To the west were seven blocks of trees, ponds, and overgrown open land before one reached Frankford Avenue. The older neighborhoods of the community stretched out from the park and Longshore Street.

Within this pattern of homes, well-established ethnic enclaves remained. The Irish were, for the most part, located in the southern end of the area to the west of St. Leo's Church. The Italians were generally in the area southeast of Cottman and Torresdale Avenues. The center for this community is Our Lady of Consolation Catholic Church at Tulip Street and Princeton Avenue. Blacks could be found in the area between the railroad and the river around Princeton Avenue and State Road. The center for this community was the Star Hope Baptist Church, now located at Friendship and Hegerman Streets. There were about twenty Jewish families in Tacony, all living above stores along Torresdale Avenue.

The Protestant community is best viewed by examining the churches of the community. The most financially well-to-do Taconyites were Episcopalians or Presbyterians. The areas adjacent to these two churches had the largest and most expensive homes. The Lutheran church conducted services in German and was located on the western edge of the community. Many of its members were farmers from outside Tacony whose families had been members long before Disston came. The Methodist and Baptist churches were (and are) located in the center of town with their populations widely spread across the community. In essence Tacony was an ethnic and religious miniature of the city of Philadelphia.[52]

The schools of the community were the St. Leo Catholic Grade School and the Mary and Hamilton Disston Elementary Schools. At 8:30 each morning one could see the lines of children walking east on Unruh Street on their way to St. Leo's, while a similar group headed west on Knorr Street to Hamilton Disston Elementary School. At lunchtime and at 3:00 P.M. these processions would be repeated. There were few snow days and no busing. Brothers, sisters, and relatives were all in the same school. The teachers lived in the neighborhood and knew the children and parents outside of school. The number of students in each grade was so small that homogeneous grouping of the students was not possible. All children were taught together, from the brightest to those needing help. Education in those days was a community matter.

After finishing the eighth grade, Tacony children either went to St. Hubert's High School (predominantly German Catholic girls), Northeast Catholic High School (boys), or Harding Junior High School. After Harding, public school students could elect to attend Frankford High School (boys and girls) at Oxford Avenue and Wakling Streets, or Northeast High School (boys) at Eighth Street and Lehigh Avenue in Kensington. All these schools, with the exception of St. Hubert's, required Tacony children to travel far outside their neighborhood.[53]

These great distances prompted the executives of the Mary Disston Estate in 1936 to donate land southwest of Hamilton Disston Elementary School for a proposed Jacob Disston High School. However, a shortage of money at the school district, caused by the depression, prevented the new school from being built. The land was eventually passed on to the city and is now the site of the Vogt Recreation Center.[54]

Safety and police matters were handled by the Twenty-seventh Police District. At first the district provided the justice system for the entire

Northeast, with horsemen starting out each morning up Cottman Street to the Montgomery County line and returning each afternoon. With the advent of the automobile, the police patroled Cottman Street in cars.[55]

Local merchants on Torresdale Avenue had their own policeman assigned. He walked a beat and surveyed the neighborhood each day and was on a first-name basis with the children and shopkeepers and could spot a troublemaker. The shops and streets had no graffiti or vandalism.

The most striking feature of the community was the lack of saloons and the closeness of the families. There were few town or factory secrets: If someone was running around, everyone knew it; if someone was a heavy drinker, everyone knew it. Teachers knew the parents of the children, in many cases having previously taught them.

Transportation to and from the community was provided by the 56 Trolley, which traveled along Torresdale Avenue. No longer did boats traverse the Delaware, or the trains carry large numbers of local commuters. Shopping was done locally, except for rare all-day trips "downtown" to Gimbel's, Lit Brothers, and Wanamaker's for a special sale. The local stores were between Knorr and Tyson Streets on Torresdale Avenue, clustered around the Liberty movie house, which had been transplanted from its original location on Longshore Street. The expression "going up the Avenue" to a Taconyite meant going to these shops. The town no longer centered on the Disston factory, the Delaware River, and the Longshore stores; now it centered itself on the 56 Trolley line and the Liberty.

The movie house was the community's main means of entertainment. Saturday matinees were packed with children gathered to see their favorites—the Bowery Boys, Bob Hope and Bing Crosby on the road to somewhere, John Wayne, and Hopalong Cassidy. Saturday and Sunday evenings also had sellout crowds.

On Memorial Day there was always a parade led by the American Legion William D. Oxley Post. Boy Scout troops led by Tacony's first troop, No. 24, were followed by a band and the men of the post. The same Oxley Post raised money with a minstrel show at Lincoln High School during the 1950s. As with the "Disston Minstrels" of earlier days, malapropisms, racial slurs, and blackface were part of the show, which featured war veterans. Such shows were an unquestioned Tacony tradition. The Disston Playground held a full day of activities on the Fourth of July, just as it did forty years before.

World War II was marked by increased production at the Disston plant. Workers with specialized skills were deferred from the armed services as the plant went on a three-shift, twenty-four-hour day. Because of labor shortages, some men worked two shifts, plus Saturday and

Sunday. The boys at the front needed Disston's steel, and the men and women of Tacony would produce it.

Signs of the war were everywhere in those days. Houses displayed the serviceman star in the window to let passersby know that the family had lost a son in the conflict. Antiaircraft guns were placed in Tacony park with a platoon of soldiers to protect the Tacony-Palmyra Bridge. Gasoline was rationed, along with butter and sugar. Victory gardens were everywhere, and no one discarded fat or rubber tires—they were recycled for the "troops Over There."[56]

The most notable local enlistee was Al Schmid, a worker from nearby Dodge Steel who became a Marine. On August 21, 1942, during the Guadalcanal campaign in the Solomon Islands, Schmid's single-handed action held off four hundred Japanese in the Battle of Tenaru. Schmid was awarded the Navy Cross: "For heroism in the line of his profession as a machine gunner. Schmid's machine gun squad was attacked by the enemy. Lacking protection from rifle men, it was necessary to tear down their frontal protection in order to meet the charge of the Japanese." In the ensuing fight, all the Marines were killed except for Schmid, who was found by a relief squad still holding his machine gun with more than two hundred dead Japanese soldiers on the battleground in front of his position. Blinded by a mortar shell, Schmid had spent the night firing at sounds. He immediately became a national hero, and the movie *Pride of the Marines,* starring John Garfield, was made about his life. During the making of the movie it was a daily treat for Taconyites to seek out Garfield and the movie crew from Hollywood as scenes from the movie were shot at Schmid's home on the 6500 block of Tulip Street.[57]

All of Philadelphia turned out to welcome Schmid when he returned home in 1943. Walter Annenberg of the *Philadelphia Inquirer* handed Schmid a Hero's Award and a $1,000 check in a citywide celebration held at Rayburn Plaza. Judge Vincent A. Carroll encouraged the blind hero, stating, "The light of your eyes has not dimmed because you have taught Americans to see the means for ultimate victory through sacrifice." As in World War I, Tacony had its hero—this time one who would be shared with the nation.[58]

With the end of the war in 1945, the town attempted to revert to prewar days, but somehow things were not the same. Many returning veterans spent their first four or five months after discharge at the Disston playground. For their service, they were entitled to receive $21.50 a week for six months, and most of them took their time finding work. It was not unusual to have forty men on a summer week ready to play a choose-up baseball game. There was a mentor available for every child in the neighborhood, and a whole generation of successful athletes were nurtured in Tacony's environment.

Sports teams were everywhere. The Tacony Athletic Association, under the auspices of Dan Carson, Republican leader in Tacony, operated across from the police station on State Road and Longshore Street. The team played in the Penn Del League very successfully during that time. At a lower level there were numerous independent teams, the most influential of which in Tacony was the 27th PAL under the leadership of Joseph McCloskey. Connected with the 27th Police District, the team won the National PAL Championship in 1949–1950. McCloskey became known to every Tacony boy over a thirty-year period for his work with the youth of the community.

Tacony was a good place to live—no bars, ethnic diversity without ethnic hostility, and work for everyone in town. All this made for an optimistic future. Tacony itself was a symbol of what progressive industrial thinking could create in a model community. Faith in the community rested on the belief of each member in the ability of the Disston saw works to keep everyone employed at a reasonable salary. This unquestioned faith came out of years of Disston paternalism and would be destroyed when the company was sold in 1955.[59] The organization of Disston's workforce by the AFL/CIO during the 1930s ultimately played an important role in the Disston family's decision to sell the company in 1955.

CHAPTER

8

Unionism Begins at Henry Disston & Sons

The CIO, the AFL, and Strikes, 1935–1950

DISSTON WORKERS were never so awed by their employer's benevolence that they lost interest in better wages, hours, and working conditions. Yet the company discouraged disruptive forces like unionization and ethnic rivalry, but a workforce sharply divided between the interests of skilled workers and nonskilled workers doomed the company's efforts to keep peace between the workers. At Disston, unionization finally emerged out of the conflict between unskilled and skilled saw-makers and steel-makers.[1]

Skilled workers were always treated well at Disston, based on their knowledge and skill of steel- and saw-making techniques. During the last quarter of the nineteenth century, native-born and Northern European skilled workers dominated the steel trades and cooperated in imposing their own standards of wages, hours, and work rules through their craft union, the Amalgamated Association of Iron & Steel Workers. The union became less important nationally as mechanization made the workers' skills increasingly obsolete by 1920. Steel companies could now hire nonunion, unskilled, Eastern and Southern Europeans to do more of the work. Beginning in 1930 the skilled steel-makers who were unionized began to lose their advantage as the depression forced people out of work. By 1940 the excess of skilled laborers and the mechanization of the steel industry reached the point that steel-making no longer required the skills of their nineteenth-century counterparts.[2]

Even this craft union was not organized at Disston, but by the 1930s a new kind of unionization that focused on rank-and-file workers, not the skilled worker, emerged. The Congress of Industrial Organizations (CIO) became the organization the workers turned to when Disston's industrial welfare system and the organizations in their communities failed them in the Great Depression. The new unionism promoted a new class consciousness that united workers across race, ethnicity, region, age, and sex, and made common action possible. Add to this the unifying forces of mass culture, and the acceptance of the union movement is more readily understood.

Workers in Tacony shared experiences such as going to the movies at the Liberty theater on Torresdale Avenue or listening to the same radio shows—experiences that tended to make them think as one. Heroes and villains were easily identified in these early, simplistic dramatizations of the human experience. It was easy to make the bosses and owners the villains in a complex business world that was not easily understood by the workers. Shopping at the newly opened Great Atlantic & Pacific Tea Company, a grocery chain store opened on Torresdale Avenue, not small ethnic stores, also brought a more national product to the community and a more common shopping experience. Advertisers and marketers learned to make distinctions within the "mass market," gearing products to a new working-class market. This in turn galvanized workers in a common understanding about what was necessary and important in the new, modern, industrial world. It also gave them an opportunity to think about their community in a different sense. Rather than being the town of Henry Disston & Sons it was becoming a town where the people were engaging more and more in discussions that were not connected to the factory.[3]

Rather than depoliticizing workers, such experiences rapidly changed mass culture, enabling the workers to grasp the significance of unity between groups and allowing them to move away from the cultural fragmentation caused by ethnic pluralism. Mass culture tended to increase political awareness as workers strove increasingly to become part of a newly emerging middle class. Captured by the belief that Franklin D. Roosevelt was willing to try new ways of dealing with the depression, workers reevaluated their fragmented political voting patterns. In Tacony this political awareness was manifested in the election of Tacony's first Democratic congressman.[4]

Tacony politics in the decades before 1932 centered on the 23rd Republican Committee, housed at the Tacony Club. Two important figures during this period were Deputy Sheriff Dick Seed and Councilman (later Congressman) Peter E. Costello. Seed, a star athlete, the left-handed member of the famous Seed brothers pitching team, remained popular in

Tacony throughout his lifetime. Unlike his brother Billy, who in later years became a dissolute character, Dick was a respected figure whom Taconyites sought out when they needed a political favor or a job.

While Dick Seed was important to Taconyites locally, Peter E. Costello was by far Tacony's most influential politician between 1890 and 1920. From the days of Hamilton Disston, the firm and the family supported Costello and the Republicans. He was a Disston worker, a Tacony builder, and the developer of the Holmesburg, Tacony & Frankford Railroad Company in 1901. Elected in 1894 to Common Council from the 41st Ward, Costello promptly became chairman of the Committee on Finance and the Committee on Highways and Survey. His political ally was State Senator "Sunny" Jim McNichol. In 1903 Costello was appointed by Mayor John Weaver to the post of director of public works, and he supervised the building of the Torresdale Water Works next to Tacony. In 1905 he was reelected to Common Council and remained in this post until 1914, when he was elected U.S. Representative from the 5th District. In Congress he served on the Rivers and Harbor Committee until 1922. In addition, Costello was vice president of the Millard Construction Company, which built the northern extension of the Market Street subway.

As councilman, Costello introduced ordinances for construction of the elevated train to Bridge Street, the extension of the Roosevelt Boulevard, and the opening up of the northeast section of Philadelphia. His son, John N. Costello, followed him to Congress, serving Tacony until 1931. These Republican Taconyites, and their Tacony constituency, were instrumental in the campaign to develop the city's northeastern quarter.[5]

In 1932 the Great Depression brought a new brand of politician to Tacony. For the first time, a viable Democrat was put up by the workers of the community. Frank Dorsey had been born and raised in Tacony and worked from boyhood at Henry Disston & Sons. The Disstons appreciated Dorsey's ability and his faithfulness to the firm. Despite their continued support of Republican candidates, they could accept the idea that a "Disston man" might be a Democrat. The Dorsey family was well known to Disston workers and the Tacony Community since Frank supervised hiring at Disston, and his brother Harry had been personnel manager. Both men had been part of the Tacony regiment that went to France in World War I. Frank Dorsey was a friend of the workers, a popular athlete, and a graduate of St. Leo's Catholic School. Dorsey's platform was modeled after that of Franklin D. Roosevelt, who promised bold moves to get the nation out of the depression. Dorsey was elected in 1932 and reelected in 1934. The workforce of Tacony no longer was aligned with the political preference of the Disston family. This political change heralded a shift in the attitude of Disston workers toward the company.[6]

Wage cuts and work slowdown caused by the depression in the years between 1930 and 1932 gave the CIO labor union fertile ground at Disston. Open talk of organizing workers surfaced there for the first time, especially among the unskilled laborers. However, the process of unionization would take years, because skilled Disston workers whose families had served the firm for generations were not as interested in the union movement. To them, the Disston family were doing the best they could to provide work. But to those who elected Democratic Congressman Frank Dorsey, it was time for the company to change.[7]

The circular-saw anvil men, who held and positioned the circular saws for the smithers, were the first to strike in 1932 over reduced wages and piecework. Following a wage cut of 10 percent the year before, a new 16 percent cut had just been announced. Without even forming a union, they simply refused en masse to report to work. The smithers supported their co-workers, and all saw work stopped. The strike lasted a month. Finally, William Klotz, superintendent of the circular-saw shop, negotiated a settlement. All workers received a 75-cent-an-hour increase with no more piecework. The company set aside its original demand that strikers return to work or be fired, and a new letter from plant superintendent Dick Nalle granted them full amnesty, declaring that if they returned to work "the past would be forgotten." The work stoppage in the smithers' shop and a generalized movement to organize the workforce prompted the company to grant a pay raise of 10 percent in mid-February 1933.[8]

In 1935 Disston workers affiliated officially with the CIO's Amalgamated Iron, Steel & Tin Plate Workers Local 1073, the traditional craft union of steel workers. The transition was peaceful until the newly formed CIO recruited the Amalgamated Steel Workers as one of its affiliates. Hostilities began nationally in 1936 when a new organization known as Labor's Non-Partisan League (led by George Berry and supported by John L. Lewis of the CIO) actively worked for the election of Franklin D. Roosevelt. The AFL refused to join the league because such a political movement was not in keeping with the AFL's nonpartisan political stand. Lurking beneath the surface was the fact that the CIO was just forming and in direct competition with the AFL membership drives. In 1937 the AFL declared open warfare by passing a resolution at its 1937 convention that opposed every political candidate "who would in any way favor, encourage or support the CIO."[9]

This competition between the AFL and CIO changed the structure of the old Amalgamated Iron, Steel & Tin Plate Workers. The council of the CIO decided to seek new members aggressively through a newly organized Steel Workers Organizing Committee (SWOC). This committee supported a drive to solicit Disston workers for the CIO.[10]

On March 19, 1937, the CIO, representing the Disston workers, signed a collective bargaining contract with the company negotiated by Edward J. Lever, field director for the Steel Workers. There was never a threat of a strike during the three weeks of talks leading to the agreement. The new contract provided for an eight-hour day, a forty-hour week, seniority rights, and an impartial chairman to settle disputes. It required two pages to cover all contingencies.

The struggle for union supremacy was played out between the employees at Disston & Sons between 1936 and the time the company was sold in 1955. It was not until 1949 that a common hatred of the newly enacted Taft-Hartley Act encouraged a national merger in 1954. The newly created AFL-CIO finally brought internal labor peace to Disston & Sons.[11]

Nevertheless, work remained slow and wages low, and despite efforts by foremen to give a little work to everyone and avoid layoffs, dissatisfaction grew among workers, especially those in the file and saw divisions. The union used these conditions to advance its cause. Disston management reacted with an occasional firing on the grounds of overzealous recruiting, but the agitation for membership continued. The union's critical selling-point was the low wages. Steady work had, in the long run, always compensated for low wages. Yet the Great Depression had forced many of the men at Disston onto short time or piecework, injecting uncertainty into the weekly paycheck. Once the economic slowdown had its impact on work schedules, the wage issue became more important to Disston employees.[12]

As orders for military steel and tanks increased in the plant, relations between management and workers improved. Disston held an acknowledgment ceremony for workers of long standing. Five employees were honored at a special dinner in 1938 for their sixty years of continuous employment at Disston & Sons. Pins were given to seventy-three other employees, whose service totaled 4,210 years. The large number of "old-timers" whose fathers and brothers had worked at Disston underscored the workers' exceptional loyalty to the company and the success of industrial welfare. An example was William Rowen, a smither trained by the "old-timers" in the circular-saw shop. Rowen disliked the idea of a union shop because it did not represent the best interests of all workers, and he was offended by the emphasis on salary to the detriment of pensions and medical benefits. In wartime, as more laborers were hired, the union became stronger and the demands for higher wages increased and became more generalized.[13]

The increase in the military contracts in 1938 allowed the family to have dividends and the workers to have a raise. Management offered the membership two wage increases amounting to 15 cents an hour effective August 1 and 18, 1938, but it cut back on the money being placed in the pension fund voluntarily.[14]

During the next negotiations in June 1939, the election for representation was again won by the CIO's Amalgamated Iron, Steel & Tin Plate Workers Local 1073. As the title implies, the majority of this union's members were located in the plant's steel mill. Wage increases of 10 cents an hour were granted effective July 1, and employees who claimed to have been unfairly dismissed were rehired.[15]

Despite these few minor union disputes, the company emerged from the 1930s without severe labor problems. *Time* magazine broadcast to the nation in 1940 that "the ancient and honorable firm of Henry Disston and Sons Inc." was "America's No. 1 sawmaker" and that Henry Disston's skewed-back saw was the favorite of American workers. The firm made more than five million saws and blades a year and controlled 75 percent of the U.S. handsaw business. Its products varied from a tiny jeweler's bandsaw blade (thickness .005 inch) with eighty-eight teeth to the inch, to a ten-foot, spiral inserted-tooth monster used for lumber and metal-cutting. Disston knives, files, and other tools were sold throughout the world. The net worth of the company was assayed at $8.5 million.[16]

However, when the company opened negotiations with the union in February 1939, it started a series of events that would finally bring labor peace. Despite an expiration date of February 28, 1939, discussions dragged on, and a strike date was set by the union for March 7, 1939. Michael Harris, subregional director of the SWOC, requested that a federal commissioner be called from Washington to mediate the situation. On March 3, P. W. Chappell arrived in the city and met with Vice President Richard T. Nalle, representing the company, and Anthony Wahner, representing the union. Nalle demanded a no-strike clause and the right to install new machines that altered working conditions without the union's invoking arbitration to decide questions of wages. Wahner wanted five days of vacation with pay instead of the current two-day provision.

Chappell reported back to Washington that these issues were on the table but that a number of issues loomed large beneath the surface. The union had recently dwindled in membership, and a vigorous membership campaign had been organized by the CIO's SWOC: "Reading between the lines it was easy to see that promises had been made during the drive which the negotiating committee found it impossible to make good." Chappell reported that the consequence of the drive was that the union took every opportunity to have heated arguments over the "phraseology" of the company's offer and had so worked on the membership that they were likely to get out of hand.[17]

In his search for a middle ground, Commissioner Chappell impressed on the union that the issues on the table did not merit putting 1,700 people on the street. Wahner finally agreed that if the company would give the five days of paid vacation and eliminate the no-strike clause, he would handle the other matters with his membership. Later the same day, Chappell told Nalle that if both sides concentrated on the two basic issues and avoided wrangling over phraseology, the strike could be avoided. Nalle agreed, commenting that the previous negotiations had been carried on for the union by Ed Lever, a man of considerable experience in labor matters, whereas Harris lacked such experience and caused more dissension than agreement between parties. Nalle agreed that the company would be willing to fund a five-day paid vacation for the workers as long as it had some assurance from the union that there would be no demand for a pay raise in the near future. Chappell suggested to Nalle that a new proposal might be written offering the five vacation days and, without using the words "no-strike provision," providing for no pay raises in the immediate future. After considerable discussion, Nalle agreed to rewrite the proposal and allow it to be submitted to the union at a meeting two days hence. He assured Chappell "that he would not squabble over words but would confine himself to essentials."[18]

That evening Chappell met with Michael Harris, reviewed the situation with him, and arranged for him to get his committee together the next day for discussions. Harris expressed concern over Chappell's actions, worrying that "perhaps he was premature in asking [Chappell] to enter the case." Chappell assured Harris that, on the contrary, the visit to Nalle "had been at the right psychological time to influence him into giving more than he would otherwise had been willing to give." Chappell's keen insights and mediation skills resulted in a contract signed on March 8, 1939, thus averting a strike.[19]

Despite the agreement, a dispute arose on November 3, 1939. After a time study of the hourly and piecework rates of pay, forty-five machinists were reclassified, some with the same wages, others with wages increased or decreased. The workers were opposed to the methods used by the time-study engineer, walked off their jobs, and demanded wage increases of 10 percent.

Commissioner B. P. Holcombe, assigned to mediate a settlement, met with Nalle, who agreed to advance the rates of pay slightly over those established by the evaluating engineer. Strikers returned to work on November 27, 1939, when these adjustments were made.

In March 1940 two thousand Disston employees again had a chance to vote for union representation. The unions vying for control were the incumbent CIO and the newly organized United Saw, File, and Steel Products Workers of the American Federation of Labor (AFL). Bitter feelings were aroused as the two groups struggled for control of the Disston workforce. The election had to be held away from the work site—at the Tacony Athletic Association—because the two sides could not agree on a location within the plant.[20]

Smither William Rowen had just been appointed foreman when the union election of 1940 took place. Talk about a strike dated back to the first union election in 1937, but none had materialized. Like most older skilled workers, Rowen did not like the union or its tactics. In his opinion, the unions did little more than cause dissension among the workers. Rowen found AFL organizer Walter "Reds" Norton particularly abrasive. He was a loud talker whose tactics, in Rowen's view, accomplished little beyond making his own job as foreman more difficult. However, men like Norton were responsible for the AFL's election victory (1,096 for the AFL versus 723 for the CIO affiliate and 40 for no union). Even after the election was over, the workers in the shop continued to bicker over representation. The older CIO union worked to increase its membership for the next affiliation election. This put the newly elected AFL on the defensive and made its negotiations for a new agreement crucial to its continued tenure in office.[21]

Norton, who lived in Bridesburg, began to talk to Rowen's workers about a strike, trying to persuade them to adopt the 1932 tactics of the anvil workers and quit their jobs. Rowen and other workers objected that such a work stoppage would be considered a strike and that those who left would lose their pension benefits. The agitation continued, with the Disston management aware of such discussions but unable to do much about union activity.[22]

On April 2, 1940, the National Labor Relations Board certified the AFL union as the bargaining agent for the company's employees. This formally ended three years of CIO leadership under an open shop. Yet the company still had more than two hundred nonunion and paid members from the rival CIO, which vowed to recapture the leadership from the AFL at the next election.[23]

The nonunion employees at Disston tended to have the same characteristics. They were usually from Tacony, longtime employees, skilled workers, and local community church members. These workers sided with the less aggressive CIO. Foremen were also opposed to the new AFL because their forceful defense of workers threatened their authority. All those workers favored the CIO as the lesser of two evils. It was clear that the AFL victory did not unite Disston workers. The aggressive leadership of the AFL now attempted to persuade the workers to join their union. If this failed they could always use coercion.[24]

With the question of representation unsettled, how would the new negotiating committee for the February 1941 contract renewal be organized? A letter sent by the AFL leadership to the Disston management in mid-April 1940 set in motion a chain of events that would greatly affect union/management relationships for the next two decades.

In an effort to solidify their ability to represent the workers at Disston, the newly certified AFL demanded a closed shop (that is, compulsory membership in their organization) and immediate consideration of increased wages. On April 26, 1940, came the response from management. A union shop was not acceptable to the Disstons. They maintained "that it made no difference to them whether or not employees joined a union or what union" and that forcing membership in a particular union on all employees, regardless of their preference, was undemocratic. Although management had no objections to a union shop, it refused to force employees to pay a fee in the form of dues as a condition of work at Disston. The choice should be left to the individual.[25]

As for the wages, the company responded that the union demands "are entirely beyond the financial ability of the company to consider." If the wage scale requested by the union had been put into effect in the 1939 contract, "the company would have operated at a substantial loss." The company reminded the union that the wage scale study that caused the walkout of November 1939 had resulted in increases in pay for Disston workers that ranged from 1.1 to 4.1 percent, with an average increase of 2.8 percent. The company promised to continue adjusting rates in accordance with the survey results, with a special department created to conduct continuous evaluations of pay rates to give better control over salary inequities. However, the company warned the union that it had no intention of negotiating in February 1941 a contract providing a basic, across-the-board wage increase. Despite the increased use of this technique in other industries, the Disstons maintained that an overall percentage increase for the entire company would erode the salary differential between the factory's laborers and skilled workers. As might be expected, this position served to divide the workers at the plant.[26]

In response, the CIO posted in each department a letter that told of an AFL special meeting held to call a strike because the company had "so far completely failed to make good on any of its promises to the Disston workers." The CIO declared that it would not support the strike unless there was a secret ballot of all Disston workers, regardless of union affiliation. The ballots would be counted by an impartial committee. Under no circumstance would the CIO support a strike that demanded a closed shop for the AFL. Finally, before the CIO considered support of any strike, it must first have a full and complete account of all negotiation sessions. The CIO assured the workers that it was "not willing to support a strike which may be foolhardy and unnecessary and caused by incompetence of the AFL negotiating committee."[27]

Posted with each letter was a fact sheet that claimed to show how the AFL "fails to deliver goods." The sheet claimed that the newly elected union had failed to deliver on its promise for two weeks of paid vacation and a union closed shop. It reminded members that the SWOC had gotten "better contracts" and that their stewards were busy and the lodge was more active than ever. It reminded the workers that the SWOC would protect its members and that "no one should fear the AFL because of its threats—they have already been shown to be bluffs." Finally, it exhorted all workers to join the CIO. A special organization meeting was called for Saturday, May 4, 1940, at 10:00 A.M. at Van Kirk Street in Tacony.[28]

The SWOC also sought an injunction to attach the union's funds, now in the hands of the AFL leadership. Anthony Wahner, now president of the AFL union, and John Lorden, AFL representative, were forced to take a determined stand on the union-shop issue, finding it impossible to operate with almost half the workers questioning their every move. Solidarity was the issue, not wages or paid vacations. On May 1, 1940, the AFL union demanded a union shop and a 10 percent increase, or 1,875 workers would strike the plant.

In a conference with Federal Commissioner Edward C. McDonald, S. Horace Disston and Richard T. Nalle offered a 5 percent wage increase and agreed to consider a broader recognition clause, but they would not consider a contract making each worker automatically an AFL member. When the negotiations became deadlocked over this issue, the union called a strike on May 14, 1940. The union was aware of the plans for a celebration of the company's 100th anniversary and predicted that the Disston family, out of love for the company, would back down if pushed. A strike date was set for May just before the celebration date.[29]

After the meeting, McDonald reported to Washington that in all likelihood a strike would occur. The CIO's sending of literature to the workers and its injunction to attach union funds gave the AFL leadership few options. McDonald observed, "If the Union shop is not granted the Union will not be effective and they will lose their union."[30]

William Rowen and others were angry that a strike date had been set without a vote. Yet at a spontaneous mass meeting of workers, the only question asked was, "Do you want lower wages or don't you?" To Rowen, the "hot-heads had taken control of the union." Lower wages were never the issue, since both management and the union had long since privately agreed to a 5 percent raise, but the issue of a closed shop left both parties in an immovable position.[31]

The AFL had the best of the discussions because it based its argument on moral issues that workers could relate to rather than simple economic demands. The AFL stood for fairness, equity, equal pay for equal work, and an end to the favoritism shown to the skilled workers during the Great Depression layoffs. It was easy to arouse emotions with such appeals. Unskilled workers were frustrated with Disston's unreliability and decreasing interest in industrial welfare programs in the face of economic shortfalls during the early 1930s. The AFL demands for seniority and a strong policy that espoused the ability of any worker to learn any job in the factory expressed the new morality of worker equality. The message persuaded workers, even when they were offered the pay raise they requested, to stay away from their jobs and insult the Disston family on the 100th anniversary of the company.[32]

Management did not grasp the moral issue. It still believed that worker loyalty to the family would be stronger than the union movement. As we have seen, management at Disston traditionally consisted largely of family members. There were six Disston men working at the factory in 1940, most in leadership positions. Described by *Time* magazine as hardy, friendly, and prolific, they remained aloof from the labor union dispute in the plant. The chairman of the board was Henry Disston, son of Hamilton; the head of purchasing was the "sporty" William Dunlap Disston (age fifty-two), whose son, William L., was in the shops, the first fourth-generation Disston to enter the business. The president and chief operator of the firm was the balding Samuel Horace Disston, fifty-nine-year-old grandson of the founder's adopted son, Samuel. He had hardened files and sharpened saws for eleven years before the family let him out of the shops. S. Horace Disston carried in his pocket a card lettered in Chinese that he translated as "Confucius say Disston has the edge."[33]

If Disston did indeed have the edge, the family intended to show it—despite impending union conflict—with an elaborate 100th anniversary celebration. In March 1940 a committee headed by William D. Disston and consisting of D. W. Jenkins, Richard T. Nalle, C. P. Smith, James Kahlert, and Eugene B. Biemuller, was named to organize it. They immediately ordered 200,000 gummed stickers with the founder's image and the years "1840–1940" to be placed on all stationery, 575 service pins (125 for 40-year workers, 150 for 30-year workers and 300 for

20-year workers), 8,000 penknives engraved "1840—Henry Disston & Sons—1940," celluloid buttons, and automobile stickers.

Organizational meetings were held twice a week in March and April, and three days of gala celebration were planned. On Wednesday, May 22, there would be a parade and motorcade from Seventeenth Street and the Parkway to City Hall and down Market Street to the river, and from there a charter Wilson Line boat trip up the river to the plant. Lunch would be served when the boat docked at the Disston wharf in Tacony, then a tour of the plant led by Mayor Robert E. Lamberton was scheduled, and a memorial plaque would be dedicated at the athletic field. In the afternoon the plant would be open to hardware and industrial customers and other business friends in the greater Philadelphia area. On Thursday and Friday, the plant was to be open to the workers' families and friends. A sightseeing route was laid out through the plant, and Disston products were to be exhibited in a tent. Light refreshments would be served nearby.

On May 1, 1940, foremen and department managers were informed of the plans and given "Guest Card Requisitions" to distribute to their workers. Volunteers were requested to serve as tour guides. Within the next week and a half, 20,000 visitors had signed up to attend the celebration. But none of these arrangements would ever be carried out. In a move that surprised management, the AFL called a strike, and work ceased on May 14, 1940.[34]

As the men left work that last day, an incident occurred that soured management on the union for years to come. In anticipation of visits by the hardware distributors, businessmen, and community leaders, the plant had been cleaned and the bathrooms refurbished with new sinks, toilets, and floors. When the strike was called, the workers' resentment spilled over into vandalism. Angry that it took a 100th anniversary to get new bathrooms, some workers tore fixtures from the wall and destroyed some newly installed sinks and toilets. Disston management would later use this incident to explain its reluctance to fix up plant facilities, "lest they be again destroyed by the workers."[35]

On the same day the celebration was canceled, a joint conference was held in the company offices. The AFL union was ready to accept the 5 percent raise but still needed to sign up twenty more members of the CIO opposition. Believing that they had protected their union from a CIO takeover, the union leadership offered to recommend calling off the strike on May 20. The union membership, however, voted to continue the strike until every worker had joined the AFL. An organizational drive was begun by the most loyal AFL union men, who were to visit every nonunion worker within the next two days.[36]

Many members of the Disston family felt victimized by the bickering between the unions. It was a divided company, with the CIO remaining very strong in the steel plant, and the saw and file shops belonging to the AFL. Nevertheless, Disston management bore part of the responsibility for the strike for the way it managed its relationships with the union. By rejecting a closed union shop, management created the battleground on which the issue of union supremacy would be settled—management's protection of skilled workers. The resultant struggle left scars that interfered for years with the operation of the plant. A union shop might have given the union more power and exposed the company to more demands for higher wages, but the relationship between the leaders of both groups would have had the advantage of stability and the possibility of building trust.

This was the first authorized general strike in fifty-four years at Disston. *Time* magazine correctly analyzed the strike as "ostensibly for a union shop and a 10% pay raise," but beneath the surface of these demands was a maneuver "to bring recalcitrant members of the CIO Steel Workers Committee union (recently defeated in an NLRB election at the plant) into line." Many workers were glum about having the celebration canceled, but little bitterness was manifested on the picket line. As one old-timer on the line sarcastically commented, "Well, they wanted a closed shop—and they got it."[37]

During the strike, S. Horace Disston went through the shop talking with watchmen in a generally congenial frame of mind. "Not being a mind reader," as he wrote a friend, he announced on May 17 that the firm had called off the celebration. Disston & Sons would have to content itself with presenting engraved Disston D-95 masterpiece saws to Mayor Lamberton, Governor Arthur H. James, and President Franklin D. Roosevelt.[38]

In letters to his friends, Disston lamented privately that he would have to "grin and bear" the disappointment. He was encouraged by a letter from the Reverend Frederick A. Tyson, pastor of the Methodist Episcopal church, wishing for an end to the strike and reassuring Disston, "Some men of my church . . . say that the company's offer was quite fair."[39]

Yet S. Horace Disston feared for the future. Expressing the sentiments of many manufacturers of his day, he wrote to Howard C. Mull, president of the Warren Tool Corporation in Warren, Ohio: "It seems as though the most you can get out of business these days is the fun of doing it, and I fear that to pay all the money that is being spent is going to leave less and less for ourselves."[40]

True to their word, the union called off the strike on May 20, 1940, accepting a 5 percent raise and a contract that contained no provision for a closed shop. There were still some CIO holdouts in the steel mill, but everyone else was now in the AFL. Nevertheless, this split among the workers remained a factory irritant in the years that followed. Even more important to the harmony of the plant was the feeling in the Disston family that the workers had harmed the workers who had gone before them by preventing the planned 100-year celebration. Paternalism was dead, and the Disston family realized it.[41]

The lack of union solidarity continued to plague company-union negotiations throughout the war years. As much as the AFL tried, they could not get 100 percent membership at Disston. Not only were some employees loyal to the older CIO, but confusion reigned in the ranks of the company's AFL membership. The suspension of Disston Local 13 of the Boiler Makers Union from the national AFL organization when some of its men refused to pay the national dues touched off another struggle for unity within the ranks of the union, just as it was preparing to negotiate a new contract in February 1941. The issues became more confused when local Disston leader Daniel Watkins sided with the local and refused to enforce the mandatory suspension from the national organization of the non-dues-paying members of Local 13. The protest rested on the men's belief that the national organization did little for the local. Again the Disston family faced the untenable position of negotiating a contract with a union that lacked unity. W. C. Calvin, president of the International Metal Workers union and vice president of the Boiler Makers Union, intervened to support the local, threatening to close down the Disston plant unless Local 13 members were reinstated to the national organization without penalty or reprisals. To prevent further escalation of the issues, James McDevitt, president of the Pennsylvania Federation of Labor, petitioned Common Pleas Judge Gerald F. Flood to end the conflict. Flood ruled that the national organization must reinstate Local 13 without prejudice immediately and that the local must enforce the national's rules. This allowed negotiations with the company to begin in May 1941, and a month later a contract was signed granting a 10 percent raise retroactive to February 9, 1941. The agreement continued until May 1942.[42]

Patriotism negated most labor disputes at Disston & Sons during the war. A wartime labor relations board was authorized by the government to prevent strikes and settled disputes quickly. The only serious wartime conflict at Disston occurred on February 24, 1943, when a strike by five hundred workers closed the plant for one night. The workers were incensed over the firing of an employee by a foreman. Michael J. Costello, federal conciliator, was called to a meeting with management and the union to arbitrate the dispute. The men returned to work, and no public announcement followed the hearing.[43]

A second disagreement was provoked, without a walkout, when a grinding trainee program was initiated and the company selected the teachers without regard for seniority. A letter from the union's executive committee to Albert Disston on September 24, 1943, settled the matter. The union decided that "in as much as it is the policy of the union to grant the best opportunity for earning larger pay to the senior men and in as much as this is a definite opportunity to earn larger pay, we see no reason why men with less seniority should be given turns to teach new men." Albert Disston backed off, and seniority became the measure by which workers were given bounty-pay assignments as personnel trainers.[44]

The Disston chapter of the United File & Steel Products Workers of America, Federal Labor Union No. 22254, opened a headquarters near the plant at 6825 Tulip Street in 1943. Dues were set at $1 a week.

Unionism brought problems to the Disston firm that went beyond strikes. First, the requirements of every job at the factory had to be clearly delineated and written down. The average time needed to complete each job was signed off by plant manager Albert Disston and the union president. Thus, a specific amount of time was allocated for the removal of sandstones by the ten-member crew charged with that responsibility.[45]

Work standards had always existed, because every foreman and worker had a sense of how much work should be done in a day, but now written parameters were prepared and became a formal part of worker evaluations. One directive specified that it was the responsibility of the operator to notify the foreman or his assistant when a sandstone was down to "7½ inches, 6 inch butt." The failure of the operator to do this cost him any overtime for that month. The sandstones were to be taken out by the crew "at ½ inches over flange, 4 inch butt." The one-inch allowance was to give the foreman time to contact the stone gang and make arrangements for the change. Half an hour was allocated for the operator to tear out the grindstone; three-quarters of an hour for the stone gang to replace the stone, and half an hour for the operator to rig up and have the machine in operation again. Total time for a change was 105 minutes. Unavoidable holdups by the stone gang resulted in the operator's being paid at the day rate, according to the union agreement. Such specifications were worked out in shop after shop until there were written specifications for all work done at Henry Disston & Sons.[46]

A second change in regulations concerned the placement of the time clock. The workers had punched in on a time clock located at gate entrances. When the union won the right to represent the workers, management complained that it was taking the workers too long to walk from the gate to their assigned work locations. If the time clocks were placed in the individual shops, the workers would get to their work more quickly. Russell McIntyre, union secretary, saw this as another indication of management's distrust of the men.[47] Rather than improving productivity, these new regulations only decreased flexibility and fostered an "It ain't my job" mentality. Wartime demands reduced complaints from the union about violations in work requests during the patriotic years of 1941–1945. But when the war ended, this lack of flexibility would cause the company great economic hardship.[48] William L. Disston represented the family's view of the union when he wrote:

> The union pressed and won job seniority on a saw, file, and steelworks basis instead of a departmental seniority system. This agreement made it costly for the company during slow business cycles when layoffs occurred. In addition, the union resisted any changes in manufacturing methods because of the fear of reducing labor hours. It resisted use of new and more productive machine tools for the same reason. The union fought against incentive rates on piece work making increased productivity impossible. They resisted cost cutting wherever the company tried to implement it and persisted in hindering management's efforts to be more competitive in the turbulent market of the 1950s. In the last ten years of the company's ownership by the Disston family, there were three strikes involving the above issues plus continuing demands for increased hourly rates.[49]

The most critical issue confronting the union from its inception was seniority. This was predictable because it reflected a demand from a unified workforce. As early as 1937 and until the outbreak of World War II, the union grappled with this problem. Despite the presence of a personnel office, most jobs were still filled by the foreman of the various departments and went to relatives, friends, and neighbors. Internal departmentalized seniority lists often conflicted with company-wide seniority claims for similar positions. The executive board minutes are filled with seniority complaints, especially in cases where workers with similar jobs switched departments. Policy statements for the Diesel Locomotive Crane and the File and Scales Room employees found in the union papers reflect two union attempts to clarify seniority issues. In both cases the union developed seniority lists that included each worker in the department and were signed off by management. Once this agreement was established, meetings were held in the department and the list was presented to the employees. Again agreed upon, these became the official company seniority lists, used for all layoffs and some promotions.[50]

The union's executive board held to the seniority clause of the contract throughout its existence. To the union, "seniority" meant plant seniority, not department seniority. To workers like master smither William Rowen, this idea was ridiculous because it assumed that every worker at Disston could be trained for any job in a short period of time. This belief irritated skilled workers at the plant, who had taken ten to fifteen years to learn their trade. They felt misunderstood and professionally undermined. The schism between the unskilled and skilled labor forces remained deep, most unskilled workers wanting higher wages and equity in pay, and skilled workers wanting increased security. Conflicts over pensions and job differentiation followed this division of interest.

Livingston Sayre, secretary of the union, worked for Rowen in the smithers' shop and typified those who demanded strict adherence to union seniority rules. Argumentative and loud (in Rowen's view), Sayre dismissed the skilled workers' arguments with a call for equality of opportunity for all workers. Sayre negotiated an iron-clad seniority rule for union officials, who, once elected to office, could not be laid off. It was the ultimate safeguard against reprisals.[51]

Seniority as a criterion for layoffs and job placement works best when everyone is doing the same work and has the same skills. This was not the case at Disston. Difficulties centering on the maintenance of separate lists in each department, and promises to return servicemen to their original jobs, made seniority a critical and continued focus of concern to both the company and the union. A policy of veteran preference had been established after the Civil War and was continued after World War I. But now two or three workers might return for the same job. The question always reverted to determining seniority.

Consider this not-atypical scenario in the Hacksaw Department. The steward, Joseph Matelli, wrote the executive committee about the case of returning serviceman Charles Bircher, who had been replaced by Pennington Reid. Reid later went into the service and was replaced by Andrew Konopka. There were now two returning servicemen vying for the same metal press machine. "Who should get it?" was Matelli's question. Another letter to the executive committee read: "In the early part of forty-two Joe Crane, the original veteran on the machine, entered the service. Frank Mikusa, the second veteran was assigned this machine, having been moved over from the sandstone department to keep the white and black workers separated, and carried his seniority with him by agreement of the union and the company." Early in 1944, Frank Mikusa was called into the service, and the machine in question went to Phil

Reed. In 1945 Joe Crane returned and was given back his machine. Crane quit in July 1945, leaving this machine open. Leon Marewski, who had seniority over Mikusa, was given Crane's job. In June 1946 Frank Mikusa returned from the service and wanted the same machine back. When approached, Marewski refused, since he had seniority and Mikusa had gotten the machine during the war only because of an agreement between the union and company to segregate black workers.

Both sides were given a chance to present their case before the union's council. After some discussion, it was unanimously decided that the seniority clause should be upheld and that Leon Marewski should continue on the machine. William Leeds Disston, one of the few surviving family members who met with union officials, recalled the difficulty facing the family when dealing with the different factions in the union. Often an apparently settled issue would be reopened in the next meeting because of a faction fight within the union. Disston management's refusal to consider a union shop in 1940 and thereafter hindered the union in its ability to control its members. In the long run, this was one of the factors that led to the sale of the company.[52]

Seniority disputes were not, of course, the main result of the war at Disston. Two rush orders for $1,250,000 worth of armor plate, one earmarked for the seats of fighter planes in France, had been held up by the strike of May 1940. Also interrupted were U.S. army orders for gun shields, light armored scout cars, and light armor plate. Short workdays and layoffs were a thing of the past. As in 1917, war had increased the need for workers at the plant.[53]

Unionism in postwar America was dominated by the battle for supremacy between the AFL and the CIO. While the official merger of these two powerful organizations did not take place until 1955, the resentment against the Taft-Hartley Act and the 1948 victory of Harry Truman with labor's help did much to end the infighting. A united union at Disston & Sons increased the pressure for higher wages, more vacation time, and greater fringe benefits. Henry Disston & Sons was ill equipped to meet these challenges.[54]

The confrontational relationship between management and workers exhibited in the depression years and into 1940 had been suppressed by wartime patriotism, but declining profits in the postwar era renewed old tensions. Gone were the days when most of the workers implicitly trusted the Disston family. The distance between worker and owner was too great to sustain the mutual understanding and sympathetic view of management that had characterized that relationship between 1880 and 1930. A labor contract that took up one page in 1935 now filled a book, and in those pages were an infinite number of items over which both parties could disagree.[55]

The Disston family was unhappy with its development, "old-time workers" resented some of the union's demands, but union members believed that the union was not only necessary for them but also could contribute to the firm's profits and productivity.

Union leader Russell McIntyre, who began working at Disston in 1941 and within a week was running four file-cutting machines, believed that Disston's lengthy apprenticeship requirements were unnecessary and wasteful of personnel. In addition, they undermined the union's cherished demands for "plant seniority." The union's position was that any worker could learn any job at the factory and that therefore a worker dropped from one job, if he had seniority, could be quickly trained to replace another worker in another department. For this reason the union advocated a three-week training period implemented by skilled workers receiving overtime pay. The apprentice period was shortened, and union trainers got extra pay. The company bought the plan because it meant full-time production from the regular worker and trainer sooner.

McIntyre's opinions were shared by the most moderate factions of the union. Not a zealot, he had joined the union because he apprenticed under a union man. His ability to keep records allowed him to rise to the position of union secretary/treasurer, a post he held from 1973 until he retired in 1988. McIntyre had the greatest respect for the Disston family and clung to his faith in the family even in retirement. The belief among the workers in 1941 was that if "Disston gave you a job, it might not be top dollar, but it was employment for life." Faith in the family's benevolence persisted; McIntyre believed the Disstons "never laid off a man if they could avoid it." Workers skilled at machines would be given windows to clean, floors to sweep, stock to move, or some other task. Workmanship was the key to a good employee. "You [the person closest to the product] were the quality control person!" The workers "had a say in the work they did." However, it was the union's job to seek higher salaries and seniority rights for everyone it represented.[56]

In May 1945 the issue of a union shop was ostensibly reconciled between the company and union boss Peter Hyslop. This tenuous agreement was contingent on the union's establishing a system that would halt wildcat strikes—unauthorized stoppages and walkouts. But to the union's embarrassment, the Steel Department was even then getting ready to walk out, which would have negated the closed-shop agreement before it was signed. Hyslop wrote to Livingston Sayre, chairman of the union's executive board, to ask him to intervene.

The problem was that the chippers, workers who chipped steel as part of the shaping process, had agreed to a bonus contingency under which they were guaranteed a 15 percent bounty above their base rate over an eight-week period. This amounted to about an extra 30 cents an hour for a total wage of about $1.18 an hour. They seldom fell below the standard quota that would activate the bonus arrangement.

Occasionally, however, they got some tough steel to chip, which prevented them from making the bounty. The company offered to compensate them at 50 percent of extra weight for time consumed in the extra work and the loss of the bounty pay. The chippers in the Steel Department refused the offer and threatened a wildcat strike. Hyslop reasoned with them, explaining that they did not have a legitimate grievance because the union had agreed to the settlement and that their strike would hurt the sensitive closed-shop negotiations with Disston management. The chippers refused to listen, forcing Hyslop to write Sayre, "Perhaps we can relieve this situation without too much confusion or getting the entire membership worked up over it."[57]

Although Hyslop ultimately managed to quiet the chippers, worker unrest continued over the issue of higher wages. This concern precipitated a nineteen-day strike in March 1947, when the company's offer of a 6-cent-an-hour wage increase to Disston's 2,500 employees was rejected by the members of Local 22254 of the United Saw, File & Steel Products Workers of the AFL. The union demanded a 30-cent-an-hour raise. The astronomical difference between the offer and the demand was a clear indication of just how far apart the two sides had drifted. This was the first stoppage of work at the plant since the one-night strike of 1943, and it was the longest strike since 1940. Mass picketing in front of the main gates prevented office workers from entering the plant on March 22, 1947. A 10-cent-an-hour raise, agreed on under the Federal Mediation Service, led by John R. Murray, settled the strike. Charlie Walker, president of the local, and John Entwisle, a Disston vice president, signed the contract and the workers returned to their jobs.[58]

Backed by contract, new work rules, seniority rights, and lessened power for the old-timers, a unified AFL union was in place at Disston & Sons by 1950. The workers now appealed to the union for help in most matters, distancing themselves from a less and less interested Disston family. For the family, the workers' strike of 1940 proved them unworthy of the kind of benevolence practiced by their ancestors. Family members in management positions no longer found the business fun because the hostility of the workers seemed to grow with each passing year. The existence of a union became verified proof to the family that businesses in post–World War II America would face continued antagonistic relationships with their workers. It was a world they did not understand, and one from which they would eventually retreat.

CHAPTER

9

The Selling of Henry Disston & Sons:

H. K. Porter Company Comes to Tacony, 1950–1984

WHEN WORLD WAR II ENDED, twenty-four-hour production continued at Disston, even though the outworn and outmoded equipment increased the cost of the final product. The worldwide market offered American firms no competition. The total destruction of the European and Asian industrial complexes left the United States with the only functioning, viable steel plants in the world. With raw steel products from America in tremendous demand, Disston's obsolete, worn-out equipment could still produce a profit. This decision to gain short-term profits would lead to long-term disaster. When the modernized steel plants of Germany, Japan, and England began production between 1949 and 1952, these profits soon disappeared. Future economic superiority would rest on the efficiency of new electric furnaces, shorter production lines, aggressive leadership, and a substitution of plastic for steel in certain manufacturing processes.[1]

It is instructive to compare the postwar Henry Disston Inc. with another Tacony firm. Unlike Disston, John Nesbitt's Company, a small manufacturer of air heaters and air conditioners, was preparing for postwar expansion. Nesbitt's had moved from Atlantic City in 1929 and remained a small company until wartime contracts for the manufacture of military goods gave it capital and resources for expansion. It positioned itself well for the postwar building boom and the explosion of the air-conditioning industry. Its buildings were in good repair, clean, and modern, and the industrial machinery was in excellent shape, making the plant far more inviting than the much older and overused Disston facilities.

Once air-conditioners became a common household appliance, Nesbitt's gross sales grew from $7,977,000 in 1953 to $23,783,000 in 1962. During this decade, Nesbitt hired far more men than Disston and became the number-one employer of Tacony youth. The *Nesbitt Newsbits* was filled with accounts of local Tacony and Holmesburg graduates going to work at Nesbitt. Its workforce during this period grew to about 1,800, while Disston's slipped to less than 1,000 as H. K. Porter liquidated the company's assets. Lack of foresight by Disston management helped undermine the company's economic viability.[2]

A significant factor in postwar industrial economics and the survival of family-owned companies was a new breed of "acquisition" businessmen. Caring little about the type of business they purchased, the acquisition people turned fast profits from undervalued family-owned companies that were experiencing financial difficulty. Operating on two premises—that tax advantages could be gained by combining companies that were money producers with those losing money, and that risk could be minimized by diversifying ownership—these acquisition people raided many of the family-owned firms of America.

The effects of these national trends were played out in Tacony through a series of events that eventually forced the Disston family to sell their saw works. Union demands, the change in the firm's family leadership, the arrival of acquisition man Thomas Mellon Evans, and the company's lack of success in developing power tools contributed to a fall from grace of the world's leading saw firm.

One more factor was a new conservation movement aimed at cleaning up the nation's rivers. As early as 1945 it became unlawful to discharge into any river or stream water that could not pass a potability standard suitable for human consumption. Henry Disston & Sons used large quantities of water from the Delaware River each day and returned it after use in the factory. Although many of the plant operations added no contaminants to the water, the law required that it be drinkable when returned to the river. The company built a filtration plant and rerouted all outlets not connected to the Philadelphia city sewage system. The work was completed in 1951, making Disston the first company along the river to comply—at a cost of $3 million.[3]

During the early 1950s, as orders declined, Disston management announced its intention to produce 6,000 handsaws a day, but this was not feasible with the outdated equipment in use at the plant. Inefficient 1876

Centennial steam engines still powered some of the belts in the sanding, grinding, and sawing rooms. Despite the modernization of the steel plant after World War I, the assembly-line methods in the saw and file divisions had changed little since 1900.[4]

Even if production methods had been adequately modernized, saw and tool orders alone were not sufficient to pay dividends expected by the Disston family and simultaneously meet the demands of a unionized workforce. In contrast to the 1920s and 1930s, few new products came on line in the 1940s and 1950s. Particularly annoying to the family was the continued failure to develop a one-man power chain saw. The company was still making a profit because of the ease with which it could sell steel to a rebuilding Europe, but the failure to develop a new product line would greatly handicap the firm in the mid-1950s, when steel production was renewed in Germany. At that time the decreased use of wood and steel in a postwar society that had discovered plastics meant fewer orders for all saws and tools. Layoffs followed, and work slowed down.[5]

The postwar years also brought significant changes to the company's long-standing family management patterns. William Dunlap Disston died on May 22, 1950, at his home in Chestnut Hill. Besides his outstanding service to the firm, William was active in the Tacony community, a member of the school board for five years, president of the Northeast Chamber of Commerce, director of the Union League and the Liberty Title & Trust Company, and a member of the Society of Mechanical Engineers. An active independent Republican, he had led in the development of the Northeast section of the city. A heart attack in 1937 had kept him from becoming president, a position taken by S. Horace Disston in 1939. A second heart attack in 1945 made him a part-time employee during his last five years. Nevertheless, William D. Disston's energy, innovativeness, and concern for the community and the workers were major factors in the firm's success during its most profitable years of operation.[6]

During the same time frame, another respected member of the Disston family left the company. Dick Nalle had served the firm since 1917, working closely with the workers. His greatest asset was his ability to relate to the workers and deal forthrightly with the union. Respect, cordiality, and the ability to build trust were his trademarks. In 1945 he accepted an offer to leave Disston and become president of Midvale Steel in Philadelphia. Later he became president of the Franklin Institute.[7]

William D. Disston's heart attack and Dick Nalle's departure left S. Horace Disston in control of the firm. S. Horace Disston had joined the firm around the turn of the century and became president in 1939. His personality was not one that endeared the Disston family to the workers—he was direct and to the point, and had little time for trivial conversation—but he was a perceptive, pragmatic leader who kept the company financially solvent throughout his tenure of leadership. He served as chief operating officer until 1947 and as chairman of the board until he retired in 1955. He died on January 10, 1971, at the age of eighty-nine.[8]

In 1947 Jacob Disston Jr. became president of the firm. A graduate of the University of Pennsylvania's Wharton School of Business, he had an unsuccessful career on Wall Street that ended with the depression in the 1930s, at which point he returned to the firm as an executive officer. The workers in the plant believed he did not know the saw business and was only interested in finance, which to them meant profit margins. In retrospect, that was just what the company needed, but Jacob did not have the personality or leadership qualities to change the direction of the firm.

Described by the workers as nice, friendly, but aloof, he was said to drink a little too much for his own good. He was well educated. A member of the American Newcome Society, organized to promote English traditions in America, he wrote and published *Henry Disston, 1819–1878: Pioneer Industrialist, Inventor, and Good Citizen.* However, workers resented his practice of hiring university graduates as supervisors despite their ignorance of saw-making. Master smither William Rowen commented that the "superintendents appointed by Jacob Disston Jr. were not good. They had business college backgrounds, but no practical working knowledge of saw making." Jacob Disston Jr. did not participate in the daily operation of the company or familiarize himself with plant routine, as S. Horace Disston had. He spent most of his time juggling financial matters but did little to promote industrial or manufacturing innovation.[9]

Moreover, Jacob Disston Jr. became company president at the most crucial period in the postwar American economy. Germany and Japan had rebuilt their industries with new equipment and new plants and were ready to assert themselves in a world market. This competition would decrease the need for American production and cut the price of products. This caused an unprecedented profit squeeze and a shifting of business that at best could be described as chaotic for longtime businessmen like Jacob Disston Jr. The market required a new view of production techniques and a new understanding of how the business world operated. These ideas were foreign to family firms like Disston.

Sales of $20 million a year in 1953 were not enough to keep the company from suffering substantial losses. The Disston family invited experts from Roebling Steel Company, bridge builders from Trenton, New Jersey, to analyze the situation and advise the family. The report on the condition of the plant was clear. Three-shift, twenty-four-hour operation of the steel works for nearly a decade had resulted in a "tired and outmoded" physical plant. Machinery was old and not in good repair, buildings needed replacement, and the entire plant required modernization. The equipment used to grind and treat steel had aged, and its assembly and finishing equipment was even more antiquated. These conditions perhaps made workers less diligent. On October 8, 1951, the upper half of a 200-foot brick smokestack blew up because an employee had neglected to remove excess gas before igniting the furnace. On June 29, 1954, the oil in a large hardening machine ignited. The machine was badly burned in the two-hour fire.[10]

A board of directors report of May 14, 1952, listed 95,159 shares in the company, distributed among 370 stockholders. Seventy-five percent of them were direct descendants of Henry Disston, and most owned fewer than twenty-five shares. The other stock had been given to loyal workers over the years, usually in twenty-five-share blocks. Eugene Biemuller and John Southwell were in this category. The largest stockholder was President Jacob S. Disston Jr., who along with Horace Disston and Liberty Title Company, trustee for Jacob's mother, accounted for 17,470 shares. Disston women accounted for at least 39,416 shares: Elizabeth Disston Brock, 969; Amanda Disston Carpenter, 1,773; Patricia Disston Coleman, 559; Gladys Disston, 13,983; Katherine Disston, 2,300; Pauline Disston Wanamaker, 5,824; Patricia Disston, 410; Lucy Disston Gilpin, 550; Katherine Disston Hopkins, 3,532; Dorothea Disston James, 969; Elizabeth Disston, 5,028; Ella Disston Stewart, 2,550; and Deborah Disston Swartz, 969. None of these Disston women was involved in the everyday running of the company, yet if they voted together they could control it. Family tradition was that these votes, combined with those of the president, decided the policy of the company. The financial crisis was about to set tradition aside, as the bank threatened to withhold dividends because of a loan given to develop a one-man chain saw. Many of the Disston women were living off the company's dividends and were understandably anxious about it. Jacob S. Disston Jr., in contrast, had his salary and a plant whose paper value belied its actual worth.[11]

Jacob's solution to the cash-flow problem involved cutbacks in development funds that would eventually hamper development of new products. Despite reducing the deficit, this settlement still left Jacob with little capital. With production slipping and the debt increasing each day, partly because of the continuing chain-saw fiasco, there was little hope of gaining a large contract or forging ahead with a new product line. Nor did the emergence of a new plastics industry bode well for Disston. Disston & Sons had miscalculated. While other companies like Maytag and Westinghouse planned for new products during the war so production could be started when the war ended, the Disston management had gambled that the same steel tool products would continue to be the backbone of its business. The 1930s venture to develop a portable power chain saw, and the failed effort to develop a gasoline-operated lawn mower (called by the workers "snapping turtles"), were financially draining, and the continued demand by the Disston women for dividends left no money for reinvestment in these ventures.

The portable power chain saw, considered by William Leeds Disston to have "the greatest potential for the Co. since the handsaw," became one of the most damaging factors in the family's struggle to hold on to the company. The issue not only split the family but identified the major cause of the firm's financial crisis—the unwillingness of the Disston family to decrease dividends and provide money for innovation and new products. In addition, the company's myopic view of itself as a "saw/steel business," as promulgated by the Disston men, limited product options in the consumers' world that followed the war. Saws and tools alone, the mainstays of manual labor, were not enough in a world where labor was becoming more automated and less physical.[12]

Despite the impending financial crisis, the union contract of August 18, 1950, reflected the company's willingness to settle quickly to avoid strikes. The new contract provided for a 7½ cent raise combined with improved pension benefits. The pension would be increased and would pay a minimum of $100 monthly, including social security benefits. Christmas would become a paid holiday, making a total of seven paid holidays a year.[13]

A lawsuit brought in 1952 by Acetogen further imperiled the firm's financial prospects. In 1947 Disston had entered into an agreement with Acetogen Corporation of Philadelphia to flame and grind armor plate, but in 1952 Disston annulled the contract without cause. Later Acetogen discovered that Disston had attempted to recruit twenty-eight of its thirty-two employees and had succeeded in hiring eleven, so it sued Disston for $1,987,358. The issue was never brought to court, but an out-of-court settlement cut into the company's cash-flow.[14]

Workers remained loyal to the firm, little suspecting what precarious economic conditions lay ahead. The 112 pints of blood donated in the 1954 Red Cross blood drive at the plant brought the workers' total since 1950 to 1,411. The fast settlement of the contract the same year reflected this cooperative spirit, if only for the moment. Forces behind the scene and unknown to the workers were determining the company's fate.[15]

In 1953 the men of the family, especially William L. Disston, Henry Bain (husband of Ellen Disston, daughter of Henry's son Albert), and S.

Horace Disston, now chairman of the board of directors, realized that new and energetic leadership was necessary if the firm was to survive. Jacob Disston agreed, and Henry Bain and S. Horace Disston interviewed John D. Thompson, executive vice president of Roebling Steel, for the job. Thompson became executive vice president at Disston in 1953. A year later he would become the first nonfamily member to serve as president.

One of Thompson's first acts was to place William L. Disston in charge of developing a special engine for a one-man chain saw. The new management had collectively decided that Disston had to get into the power chain-saw market or lose its competitiveness in the saw business. The difficulty in the past with making a profitable line of chain saws lay in the contract with the Kiekhaefer Company of Cedarburg, Wisconsin. Henry Disston & Sons was at the mercy of President Carl Kiekhaefer, who promised to develop an engine in the early 1940s but never lived up to his promises. It was not until 1948 that Kiekhaefer delivered the first lightweight engine so Disston could market their first one-man chain saw. But like the Disston two-man saw, the guide-rail and saw-chain design was too heavy for one person to manage without considerable strain. The Kiekhaefer engine also proved to have a defect in the fuel system. Carl Kiekhaefer denied that the problem was in the engine, alleging that the frame for the saw affected how the motor functioned. However, within the year he admitted that a defective fuel-pump diaphragm was at fault. This left Disston with only the impractical, two-man saw when the McCollough and Bland companies began producing small specialized saw engines. The saw business was in transition from manual to automated tools. The question was whether Disston could change production fast enough in this competitive atmosphere of fluctuating international prices and reduced sales.[16]

Its first opportunity came in 1952 when Guy S. Conrad and several of Kiekhaefer's key employees resigned and were available to work at Disston. Contact was made with them, and the next year was spent establishing design parameters for a totally new lightweight one-man saw. With the design completed, the company looked for a location to place a new small-engine plant that would free them from the whims and delays of Carl Kiekhaefer.[17]

The firm borrowed $3.5 million from Philadelphia National Bank specifically for the development of a small engine, with the commitment that the profits from the venture would immediately be used to repay the loan. Guy S. Conrad was hired to recruit a workforce and oversee the development of a functioning small engine that made possible a usable one-man saw. The engine was successfully tested at Disston, and dies were developed for its production. In the spring of 1953 a symposium was held in Wisconsin for the Disston employees associated with the chain-saw division. Land was purchased in Madison, Wisconsin, and William L. Disston made plans for a new engine plant. The location guaranteed Disston that Guy S. Conrad and those who worked on design could remain on the project as full-time workers. Disston management went to Wisconsin for a groundbreaking ceremony that was attended by Madison city dignitaries to celebrate the coming of the Disston/Wisconsin venture to their town. All this was known only to the top circle of Disston men.[18]

Carl Kiekhaefer, president of Kiekhaefer Corporation, learned of Disston's activities from his secretary, Rose Smiljanic. It seems that a number of Kiekhaefer employees were resigning and being hired by Disston. A loose-lipped secretary for one of them was telling other employees of her former boss's raise. Kiekhaefer was able to uncover enough information about the scheme to become alarmed that his business was about to suffer great financial loss.

> To accomplish their own manufacture, the Disston Company decided to play their cards under the table. Very slyly and subversively, their first step was to employ the confidence of a senior vice president of our corporation.
>
> This individual, namely Guy S. Conrad, was in our employ during the period that the new model development got under way. . . . [A] nest of refuge . . . is now taking form as the Disston-Wisconsin Corporation. It is to this refuge that our former chief engineer, chief layout man, chief detailer, chief tool designer, chief tester, etc. all came to roost as they successively "resigned." . . .
>
> The net result is a conspiracy on the part of the Disston Company, who were under contract with us not to purchase engines from others; Mr. Guy S. Conrad, our own Senior Vice President; Nelson Pattern Company, who has received as a vender [*sic*] more than $700,000 worth of pattern business from our corporation.
>
> The resulting conspiracy is the subject of a damage suit that may run into $2,000,000. With the magnitude of the stakes involved, the Disston Company decided to proceed very slyly and subversively to carry out the "Trojan-Horse episode." The fact that we have many loyal employees made them appear somewhat ridiculous, and much evidence has been piled up against Disston. This type of conspiracy suit has been brought to court and upheld many times in favor of the original manufacturer.[19]

In this letter, Kiekhaefer never mentioned the twelve years of unfulfilled promises and the fact that a successful small engine for the Disston chain saw was never developed. It is unclear who at Henry Disston & Sons gave the order for the Wisconsin project, but Jacob Disston Jr. was president when the decision was made, and he would pay the price for the decision during the company's reorganization in April 1954.

In the fall of 1953 the newly hired small-engine plant manager, Guy Conrad, predicted that selling 10,000 chain saws in one year would allow Disston to break even on the project. Yet Vice President of Sales Walter Gebhart predicted that sales would be no more than 8,000 a year—and "that would be a good year." William L. Disston felt that Gebhart's predictions were similar to those made throughout his tenure with the firm—unreasonably low so that the Sales Department would look good when such estimates were exceeded. Secretary and office worker Thelma Koons remembers Disston family members discussing Gebhart's miscalculations in this matter.

While the question was still under discussion by Disston's board of directors, news came from Kiekhaefer Aeromarine Motors Incorporated that a lawsuit was being filed against Disston, charging illegal employment of former Kiekhaefer executives. The previous Acetogen Corporation settlement made Disston management wary of another court action. The pending lawsuit, the pessimism of the sales department, and the record of continuous financial loss caused new CEO John Thompson to announce the discontinuation of the Madison project.[20]

This position was supported by members of the Disston family, who believed that enough money had been spent on a lengthy experiment that had lasted years and cost the company millions of dollars with no financial reward in sight. However, most of them were unaware of the clause in the agreement with Philadelphia National Bank (PNB) that called for the immediate repayment of the $3.5 million loan if the project was abandoned. The punishment for nonpayment was that the dividends normally paid stockholders would go to PNB as partial payment for the debt. All stockholders faced an uncertain financial future if the loan was not repaid. This forced every Disston family member who was living off the dividends to reconsider the sale of the company. To this day, William Disston believes that the decision to cancel the Madison project led directly to the selling of the firm.

The annual board meeting of April 1954 saw S. Horace Disston step down as chairman of the board and Jacob S. Disston Jr. assume that post. John D. Thompson became the elected president and CEO. It was the first time in the company's long history that the firm was headed by anyone outside the Disston family.

Immediately after the meeting, William L. Disston was again directed to explore the idea of a lightweight, one-man chain saw to replace the discontinued Madison project. William contacted West Bend Aluminum Company, developers of a lightweight 2½ HP gasoline two-cycle engine that perfectly fit the specifications for a one-man saw. The Oregon Chain Company was contacted for chains, and Disston itself developed the cutting bar. Within five months, six units had been completed and were operational. William L. Disston developed production schedules, costs, profit margins, and prices established for anticipated sales. It was at this point that he received notification that members of the family were about to sell their stock to H. K. Porter.[21]

William L. Disston is undoubtedly correct in holding the opinion that the loan-repayment fiasco forced the crisis at Disston in 1955, but it is doubtful that this or any one event can explain why the firm was sold. After the death of Hamilton Disston in 1894, the family pulled together and paid off the firm's debt despite the severity of the crisis. In 1955 there were just too many negative forces and too little family commitment to override them. The firm had become merely a means of producing money for the family, with little of the personal appeal and pride felt by the Disstons of the nineteenth century, who "manufactured the finest saws in the world." How could a firm with declining profits and old machinery, no new product beyond the beet knife, and diminishing interest on the part of the family owners face up to the reconstructed industries of Japan and Germany? If not this crisis, then the next would have had similar results. The issue was the lack of Disston family commitment to operate the firm.[22]

Disston & Sons became vulnerable to a man like Samuel Mellon Evans when PNB demanded repayment of its loan at a time when the firm had little available cash—and made it clear to Thompson that, unless the loan was repaid in 1955, the bank would petition the courts to discontinue the payment of dividends to stockholders. This was not welcome news for the family.[23]

Newspapers speculated that Henry Disston & Sons was about to be sold by the family. Offers were made by American Hardware Corporation of New Britain, Connecticut, the Frankford Arsenal, the U.S. government, and H. K. Porter Company of Pittsburgh. The government denied the rumored takeover bids, but nevertheless the stories persisted. William Leeds Disston tried to get American Hardware interested but met with strong resistance from Jacob Disston Jr. and Henry Bain. Only Porter's bid was brought before the Disston stockholders for consideration.[24]

John D. Thompson opened talks with C. R. Dobson, a representative of H. K. Porter owner Samuel Mellon Evans. After a twenty-minute conversation with Evans on the phone, Thompson remarked to Dobson, "I don't like your boss. What kind of a man is he?" Dobson replied, "He's a son-of-a-bitch, but he pays well." This set the tone for future dealings between Evans and Disston workers. Jacob Disston and Henry Bain supported Thompson in his negotiations with Evans, but William L. Disston opposed the deal because it undervalued what the firm was worth and in his judgment amounted to a steal. Nevertheless, the strength of the proposal was that family members would receive a tax advantage and would continue to be paid their dividends.[25]

American Hardware was disappointed that the Disston family had already made overtures to H. K. Porter without considering its offer, and believed it had not been given a fair hearing by the Disston management. American appealed to Billings and Spencer of Hartford, Connecticut, a nearby business associate, to help them be heard. Billings and Spencer, through previous dealings with Disston, owned 2,600 of Disston's 95,229 stock shares. The firm sent Roland J. Ahern to Philadelphia to stop the deal. Ahern went to court and applied for a temporary injunction to stop the sale to H. K. Porter on the grounds that Disston stockholders who were not members of the family were being "railroaded" into the deal. A hearing was set for November 13, 1955, in the courtroom of Judge John W. Lord, who was to decide if Disston's board of directors were acting "in their best business judgment when they decided to sell to the Porter Co." Ahern testified that "their decision may be in their best judgment but I think they could have gotten a better offer. It almost looks as if the deal is being railroaded through."

Lewis S. Stevens, the Disston company's counsel, asked Ahern "if he was prepared to prove fraud in the transaction" because "the rest of his argument was purely one of protesting judgment." Stevens said the company and the stockholders "might be jeopardized to the extent of approximately $1 million if a permanent injunction were issued. With no proof of fraud available—or even possible, since fraud was not the issue—Judge Lord agreed with Ahern about the price but ruled in favor of the sale.

The suit forced the Disston management to declare its financial status publicly. It was not a pretty picture. The firm reported a loss of $300,000 from January to November 1955 and a $5 million loss in 1954. Disston's total assets were reported at $17 million, with a debt of $4 million. Simple mathematics told the world that the firm was worth more than the projected $10 million offer by H. K. Porter.[26]

Yet the firm sold, because the president of H. K. Porter, Thomas Mellon Evans himself, sought out Disston & Sons, made the offer, determined how the transaction could be expedited to benefit the Disston

Trust, and pursued the sale to its November 1955 conclusion. Roland Woehr was an employee in the Disston office when Evans began negotiations. He had begun work at Henry Disston & Sons in 1946 as a draftsman. The Disston family took a liking to him and paid his tuition at Drexel Institute's night school so he could get an engineering degree. Woehr found the family to be fine, generous people, but "not attuned to new production methods and business practices," making them no match for the aggressive Evans.[27]

From the start, Evans was armed with company statistics that indicated high asset value, and he was well informed on the split in the family over the dividends. It was clear from early discussions that most family members had little idea of the assets of the company. He immediately saw an opportunity for a quick profit. If he could come up with a deal that continued dividends to the family, he could force the divided family to accept a sale price far below the company's asset value.[28]

The family had abhorred the idea of selling the company to another saw manufacturer, feeling that such a decision tarnished the Disston name and betrayed their ancestors. How would it look to future generations if the once-proud Disston firm was sold to a lesser company? However, there remained the possibility that their current Disston stocks would not pay dividends at the end of the year, seriously inconveniencing family members. Differences of opinion on the priorities of the company exacerbated ill feelings among the family members directly associated with its management. There was no single direct message being articulated that all family members could accept and abide by. Businessman Jacob Disston Jr. and the much younger William Leeds Disston were at the heart of the dispute. William saw himself the last of the real Disstons educated through internship in the factory and trained in all the processes of steel and saw-making. He, like his father, knew the workers, and they respected him and appreciated his recognition. William was fighting for the future of the company and advocated the development of new products. Jacob was the businessman who cared about dividends for the family and the solvency of the firm. He continually sided with company cutbacks, especially in the developmental departments of the firm, to balance the ledger. As a result, there were more workers available for experimentation and innovation during the Great Depression, when Disston & Sons used every conceivable method to keep skilled men employed, than there were during the prosperous years of the 1950s, when foreign competition and the money crunch hit steel companies.[29]

William could not persuade a majority of the family to keep the firm by working out a new financial package with investors from outside the city. William was particularly annoyed by Jacob's ignorance about the everyday operation of the plant. Tradition and Disston's reputation as the finest steel-makers in the world were on the line, and William fought Jacob's proposals at many meetings. They could agree on very little, much less on the sale of the company.[30]

The women of the family, many of whom were living off the dividends, were far from sanguine about the company's future. When contacted by H. K. Porter in September 1955, they quietly sold their shares, and Samuel Mellon Evans became the owner of the world's largest saw-making firm. Nevertheless some of the family disagreed with the sale and notified H. K. Porter of a minority stockholders' suit in October 1955 to give the family time to consider its options.[31]

Because the decision to sell to Evans greatly affected the future life-style and property values of the people of Tacony, it is crucial to understand the methods the acquisition men who came on the scene across America in the 1950s and 1960s used. What they did with their newly acquired companies had repercussions in the streets of the communities and in the lives of the workers. Their greed had as much to do with the deindustrialization of America as foreign manufacturing did.

Evans was a distant nephew of Pittsburgh banker-industrialist Andrew W. Mellon. His parents were not wealthy, but they managed to send Thomas to Yale, from which he graduated in 1931 with a degree in economics. Evans faithfully took a job as clerk at $100 a month in the office of William L. Mellon, head of the Gulf Oil Company. He later claimed that this association was responsible for his knowledge of finance, the foundation of his later success. William Mellon encouraged Evans to start his own company, and financed him by renting shares of common stock in Gulf for one year at 3 percent interest. The market rise of 1936–1937 generated enough cash profit to give Evans his own working capital.[32]

The capital was used to purchase shares in H. K. Porter, a run-down, virtually bankrupt steam locomotive company in Pittsburgh. Evans was buying up, at 10 to 15 cents on the dollar, a 1926 bond issue of $1 million that had gone into default in 1931. When the courts reorganized the company after bankruptcy, they handed the new company over to the bondholders, who received ten shares of new convertible preferred stock, $100 par, plus five shares of new common for every $1,000 in bonds. As the major bondholder of the company, Thomas Mellon Evans now had the power to elect himself president of H. K. Porter Company in 1940.

H. K. Porter was no prize, but Evans capitalized on his opportunity. He sold off the company's large office building and moved the office staff to a building at the plant site. He closed down unprofitable divisions of the plant and was able to show a marginal profit in his first year. World War II intervened, and the government gave H. K. Porter war contracts for steel products. The profits were now large enough to allow Evans to stockpile cash for expansion.

Evans knew the company could not completely diversify out of the locomotive business, and that survival of the company in peacetime rested on the acquisition of new businesses. He gradually converted H. K. Porter into a holding company, a nucleus for a group of industrial firms. His first purchase was Mt. Vernon Car Manufacturing Company and its subsidiary, J. P. Devine, manufacturer of chemicals. Evans had little cash on hand, so he employed some creative financing. He borrowed $2.7 million to meet the purchase price, and the next day declared a $3.5 million dividend from Mt. Vernon's assets, which was paid in cash to H. K. Porter. It also devalued all stock owned by small investors. Thus, Evans was able to buy Mt. Vernon with its own money and have full control over two companies. As a by-product of Evans's acquisitions, workers usually suffered layoffs, firings, and radical changes in working conditions and products. His ruthlessness and eagerness to show a profit in many cases left a company without a trained workforce, a shell of its former self.

To have more time to investigate new acquisitions, he hired Pittsburgh steel man C. R. Dobson to handle the daily operation of his growing empire. Evans and Dobson evolved a sophisticated chain-reaction system of building up cash from newly acquired companies and using it to purchase the next company. They looked for companies operating at a loss but possessing assets that could be quickly sold at profit. The best businesses, as far as they were concerned, were those that supplied industry and were capable of earning at least 10 percent on sales. These earnings would allow H. K. Porter to attract young managerial talent bent on making a profit. This young talent had to possess expertise in improving assembly-line speed, accelerating production, and determining which plant departments to close. Knowledge of how to make or design a particular product was not necessary, just financial reasoning.

Evans and Dobson's management system worked, for in the end the firms they purchased quickly showed an average 10 percent profit margin. This gave H. K. Porter the rollover cash to purchase Quaker Rubber Company, West Coast Rubber Company, Delta Star Electric Company, Watson Stillman Company, Connors Steel Company of Birmingham, Alloy Metal Wire, McLain Fire Brick Company, Laclede-Christy Company of St. Louis, Pioneer Rubber Mills of Pittsburgh, Riverside Metal Company of New Jersey, Carson and Sullivan Company, and A. Leschen and Sons Rope Company.[33]

Thomas Mellon Evans's decision to buy Disston was typical of his purchases of family-owned companies, with the exception that this time he would use little cash to close the deal. On November 14, 1955, the stockholders of Henry Disston & Sons voted to accept H. K. Porter's offer of $103 per share, and the minority suit was defeated in court so the deal could be finalized. The total cash needed by H. K. Porter was $1,332,000, combined with the assumption of all of Disston's debts. The total cost to Porter was $6 million to the Disston family in Porter shares and $4 million in assumed debt, both of which commitments could be satisfied by issuing H. K. Porter stock. Tax incentives made the deal attractive to the Disston family. Working out the tax angles appears to have been the key to all Evans's business transactions, and Evans himself analyzed the tax advantage offered the Disston family in a *Business Week* article.[34]

The Disston family members, who owned approximately 80 percent of the stock, gained financially through higher prices on the stock market. In the early part of 1955, Disston stock was estimated to be in the low $30 range. Evans proposed to pay Disston stockholders $65 a share in a senior stock-through exchange that involved no cash, which amounted to a trade-off of Disston stock for $6 million value of 4.25 percent cumulative voting preferred stock. The Disston deal is the first one Evans made in which the purchase was made exclusively through transfer of stock and not cash.

The exchange of nonvoting stock meant that the profits per share were tax deferred for the Disston family. If they sold their new stock in H. K. Porter, and if their Disston stock was acquired at less than $65 a share, they would have had to pay a heavy federal tax. If their cost basis was more than $65 a share, they could spread their loss over years, thus getting the maximum tax advantage.

This was important to the Disstons. When Evans took over the Disston plant, the first thing he did was pay off the loan to Philadelphia National Bank. Unfortunately for the workers, however, Evans obtained the cash for this payoff by confiscating the pension fund. The first dividend on the new Porter preferred stock was paid on February 1, 1956, providing uninterrupted dividends for Disston family members. Disston's single product line—saws—had meant that business was subject to stock market fluctuations. In contrast, H. K. Porter's diversification promoted stability for investors. Simply put, the Disston Trust was better protected with H. K. Porter than with Disston & Sons.

For Evans, the deal offered even greater advantages. H. K. Porter acquired an outstanding line of products in the saw field and added $20 million a year in sales, with no dilution of common stock and no depletion of its cash reserve. Other companies Evans owned could more efficiently make the raw steel for Disston, permitting the closing of one of Disston's lowest profit-margin departments.

Porter also received about $500,000 a year in depreciation due to the high valuation of Disston assets. Disston's losses in 1953 and 1954 offered Porter a tax loss of approximately $3.25 million, which, when applied against Porter's earnings, gave Evans large tax benefits in 1956, 1957, and 1958.

Later the Disston holdings were combined with a new purchase, that of Carlson & Sullivan Inc. of Monrovia, California, a maker of tape and steel rules that did $1 million in business a year. The new company was called Carlson Rules and Measures / Henry Disston Division. Despite the lack of interaction and connection of these two firms, they were for Thomas Mellon Evans a perfect match. Carlson was a cash-maker and Disston provided a tax shelter.[35]

Once Evans owned the company, assistant manager Clyde Cassell was ordered to sell off equipment, stock, and everything that could produce immediate cash. Factory buildings and land were put up for sale within the first six months of Porter's operation of the plant, and excess steel and machinery were sold outright. It was clear that Porter intended to make money by being a liquidator and not a producer. By these means, Cassell was able to raise the $6 million required to cement the sale, with Porter still owning three-quarters of the most usable parts of the original Disston plant.[36]

An October 6, 1956, *Business Week* article entitled "An Old Business Is Rejuvenated" told the story of Thomas Mellon Evans's economic success with the new Disston Company. H. K. Porter's accountants announced that after only one year the company that had been losing money under Disston management showed a $100,000 profit for the month of September 1956—despite a decrease of $5 million in sales caused by cutting back on product lines. A company spokesman claimed that "there might even be more paring. We're not looking for high volume. We're looking for profits."[37]

Product-line changes were not the only remedy Porter prescribed to bring Disston out of its financial doldrums. Equipment, production techniques, and management were examined carefully by Porter's staff. Operations such as the chain saw were discontinued. Even the profitable sugar beet knife business came under scrutiny. Nothing was sacred, everything was judged on one basis—profits.

The importance of the sugar beet knife business to H. K. Porter dated back to Henry Disston & Sons' development of a market for the product. Chief Engineer Bye's confidential memo to head of sales Walter Gebhart demonstrated in 1942 the value of the sugar beet knife business to Disston. The memo listed all sugar beet knife sharpening machines then functioning. There were machines in Green Bay, Wisconsin, and Findlay, Ohio, and there were eighteen in Michigan, four in Montana, fourteen in Colorado, six in Nebraska, four in Wyoming, ten in Utah, four in Idaho, and seventeen in California. With each machine containing ten to twenty knives, Disston had to produce thousands of these knives a year to service the industry. Nevertheless, there were signs that the sugar beet knife business was headed for difficulties. A Great Western Sugar Company report from Denver, Colorado, in March 1947 contained some surprising information:

> Our records indicate considerable variation from year to year in the tons of beets sliced per knife used. I am showing below averages for the years 1942 to 1946 inclusive:
>
> 1942 94 tons sliced per knife
> 1943 94 " " " "
> 1944 70 " " " "
> 1945 66 " " " "
> 1946 63 " " " "[38]

What was causing this alarming decrease? Was the steel less hard, causing it to wear sooner? These questions would eventually be raised because of increased sales of imported knives. In 1954, after having the market basically to itself, Disston began to be concerned about the intrusion of the Japanese and the Germans.[39]

Norman C. Bye visited the Seattle plant on October 31, 1958, mostly to examine knife production as it was then being threatened by the aggressive entrance of the Japanese into the knife trade. Bye reported to his superiors that the Seattle management "are still complaining about cost." It was clear that "Japanese plated knives gave equal performance to German [and American] knives but were lower in price." The Seattle management requested that Bye get them all available "reports on flame-plated knives and tri-ply hog knives," because these processes seemed to be the basis for the Japanese superiority.[40]

The Japanese continued to refine and produce even cheaper knives of high quality and were moving into Porter/Disston territory in the western part of the United States. Nevertheless, Porter/Disston sold the Japanese the knife-sharpening machines that would in the long run end their dominance in the sugar beet industry. When Japanese industries became world exporters in the 1970s, they were able to take the U.S. sugar beet knife industry from H. K. Porter/Disston. By 1970 H. K. Porter disbanded beet knife production because of a lack of profit.[41]

H. K. Porter appointed managers yearly and rotated them to help stimulate profits. The top management at Porter's Disston Division were either newcomers or Disston holdovers in new jobs. The Disston Division was originally headed by Lawrence L. Garber, a veteran Porter manager, but within the year he was appointed vice president for production at the parent plant. He was replaced in June 1956 by Eugene Salinger, assistant manager at Porter's Riverside plant. His top aide, Arthur S. Nippes, had been with the H. K. Porter organization for only a few months. Under Nippes's direction, production changeover, product shakeup, and personnel cuts accelerated. Porter, in an article in *Business Week,* claimed to have invested $500,000 for new equipment during the first two years of operation. The claim rested on six recently acquired Thompson electric grinding machines that had replaced 150 sandstones driven by belts. Nippes added that "a lot more machines will be brought to the plant." However, both William L. Disston and Russell McIntyre agree that the Thompsons were purchased in 1954–55 when the Disston family controlled the company, and that during H. K. Porter's ownership few if any machines were added. One can surmise that H. K. Porter managed the news to make it appear that it was improving machinery in the plant. Showing a profit, and improving machinery, gave the outward appearance of a company being rejuvenated. If Disston stock was ever sold publicly in the open market, such reports would be helpful to Evans.[42]

In the meantime, production continued using aged machines driven by overhead belts while workers installed compact, self-powered metalwork machines of late design around this antiquated machinery. Production flow lines were modernized and made more compact. Flow lines for hand saws formerly took each saw down 4,924 feet of production line scattered through several buildings. Under the new management, the production line traveled only 900 feet within a single building. Such changes increased profits and decreased expensive labor costs.[43]

Nevertheless, workers at Disston did not notice any improvement over Jacob Disston Jr.'s management. Supervisors appointed by H. K. Porter seemed more interested in profits than were their predecessors. In late 1955, the plant received an order for four 60-inch chisel circular saws with 2½-inch holders. William Rowen, chief smither, noticed the holders were loose and told the H. K. Porter supervisor about it. Rowen was worried that the teeth were also loose and would be thrown out of the circular saw as it rotated. Over his objections, however, the saws were ordered shipped to a Maryland factory. Within a week the saws were returned to Disston as unsatisfactory, with the report that an operator had been seriously injured when a chisel point was thrown and hit him in the chest. The company now had to produce new saws and pay for the shipping cost both ways. To Rowen and the workers in the shop, this was proof of the new company's lack of concern for workers and the incompetence of the new supervisor.[44]

Charles J. Norbeck, from the Maintenance Department, agrees with Rowen. "It was not the same as working for the Disstons. H. K. Porter did not seem to care about quality," recalls Norbeck. "It was cheaper to repair defective shears than it was to produce them right the first time." Much of his work in the Shipping Department consisted of replacing defective parts from Disston handshears. This resulted in immediate profits but held little hope for the survival of the product in the long run.[45]

Job descriptions were more clearly stated under H. K. Porter management. Pushed by the company, these job descriptions were aimed at promoting efficiency. The union accepted them as a protection against unfounded charges of "slackism." A penchant for detail left little question as to what a worker's job responsibilities were. Even the simplest jobs carried detailed descriptions. Take, for instance, Job 763, listed in 1958:

> General Job Description, Warehouse sorter. Hourly rate $1.59. Assemble items for customers and branch house packing. Get order from foreman and examine items, pick all items from same bin area before moving to next area, pile neatly in container truck, check items picked or line items back ordered. Mark order with total number of items picked. Tally and make out time card separating units as indicated on standard sheet. Move truck with carton to conveyor and place on conveyor for packing.[46]

Such instructions were a marked change from the quality-driven days of Henry Disston. Profit was best achieved, according to the Porter philosophy, if everyone followed the flow sheets as prescribed for each assembly line. No longer did the worker decide what was work of good quality, but the "people upstairs decided what was acceptable." Russell McIntyre remembers the company's apathy. When making jackets for printing presses on his Madison precision grinding machine, he could not meet the quality standards requested by the customer because the overused machines were in need of repair. Nevertheless, McIntyre's supervisor sent these jackets on to the customers, only to have them returned within the week. In skilled trades, such as saw-making, it was impossible to make the workers into assembly-line robots because there were too many variables. Disston made too many products—saws, files, or clippers—to different specifications. Machines had to be continually adjusted and checked, requiring the workers to be more than mere "watchers or feeders for a machine."[47]

The new management made other dramatic production changes. "We found," said Salinger, "we were still producing or carrying in stock some product for which there had been no orders for more than a year." The Chain Saw Department, once the Disston management's hope for the future, was the first to go. Further, a management study showed that Disston was making a profit on only $15 million of its sales. This meant that $10 million worth of sales resulted in actual losses for the company. Both Garber and Salinger concentrated on the unprofitable $10 million and ordered cutbacks in "individual items, rather than full production lines." For example, Salinger reduced saw-handle production—which had offered twenty to thirty assorted models—to nine basic types.

Two major lines were dropped: armor plate and specialty steel. Disston's armor plate business never amounted to much after 1950 because the company had given away its steel formulas during wartime and foreign countries could produce the same steel more cheaply. Its steel mill was old and decrepit, with equipment that broke down continually. "We found it would cost $3.5 million to renovate that steel mill," Salinger pointed out. "And then we'd still have had a marginal plant on our hands."

Inventory problems under Disston management resulted in having $5 million to $6 million tied up in excess raw materials. Under Salinger and H. K. Porter, Disston's inventories were cut to $1.5 million, freeing more cash for the company.

Under Salinger's leadership, Disston's production-line workers dwindled from more than 1,500 to 850. The clerical and administrative force was cut by 25 percent, and a production schedule force of 90 workers was cut in half. Responsible for running Disston under H. K. Porter's management system, Nippes concentrated on plant operations and personnel, while Salinger forced the parent company to reconsider Disston Saw's policy, sales, and finances. By 1956 Salinger had his marketing aides scouting for buyers for a future line of power tools. "We've dropped a lot of the unprofitable business," Salinger said, "but we'll get sales to cover that and more." All this window dressing did little to hide from the workers the continued decline in the quality of the workmanship and the lack of interest in employee safety.[48]

Bob Bachman, a longtime worker at Disston, saw Salinger's actions as just another sign of the general mismanagement under the H. K. Porter regime. He witnessed the firing of long-standing supervisors and their replacement with young workers without experience in the steel business.

Maintenance staffs were cut in half at a time when more crews were needed to replace motors and keep the heavy machinery running. One of their duties was to check the cable in the Cleveland crane, which held the ladle filled with molten steel over the mold, allowing the extreme heat to attack the cable. It took two workers two hours to replace one of the three-quarter-inch cables. This was normally done once a month and more often if necessary.

Bachman recalls one morning when there were only three workers on maintenance duty and he was floor boss at the time. The workers on the ground wore goggles as they guided the ladle into the mold. Charles Diemer was on the stage, behind the steel plates, directing the operation. All of a sudden the cable broke and the steel splashed to the floor, hitting Diemer and another worker. Flames shot up to the ceiling as the heat of the steel ignited the ground. Bachman was ten feet away from the spill. He rushed to put out the flames covering Diemer and the other worker, both of whom died later in the hospital. Most of the workers believed the accident was caused by poor maintenance, but no hearing was held, and the steel plant was summarily closed three weeks later by H. K. Porter. Bachman was retained beyond the closing date to take inventory of all the steel left on the premises, which took him two weeks. On the day the inventory was completed, he was given his check, and the steel plant closed. The excess steel was later sold to Stave Brothers Junk Yard in Philadelphia.[49]

Smither William Rowen was equally dismayed by the cuts in maintenance and shop workers. When saw-teeth foreman Sam Freas retired, the new boss caused havoc in that shop. Sam Freas had built and perfected some of the machinery for producing chisel-point circular saws and replaceable teeth. His black books, which provided milling instructions for the repair and replacement of machine parts, were thrown out when the new supervisor cleaned out Sam's desk. Without the instructions, the machines could not be repaired. On another occasion, a supervisor from Porter's Pittsburgh plant visited Disston and observed Rowen using a pinning hammer to draw up the steel in a circular saw. Believing that the marks made in tensioning were harmful to the saw, the visitor asked him to stop. Rowen had to tell him the marks were part of the process and would be removed during the grinding of the saw. These incidents and others like them convinced the workers that the Porter management did not know the saw business and cared little about learning it.[50]

Publicly, Samuel Mellon Evans reiterated that he was out to improve every acquisition. In a 1955 speech he announced: "We have found it is easier and less expensive to buy a company and to develop its potential than to start something brand new."

Explaining why family firms were his main purchases, Evans said, "How much quicker and usually more economical it is to purchase a concern in a given field and thus secure overnight its engineering staff, technical knowledge, plant, products, and customers. As all of us know, there are major tax problems involved with inheritances and diversification of investments." Evans had outlined the scheme of most acquisition men: find a tax loophole and buy the company, publicly proclaim that you are developing the firm, sell off equipment and operations to produce short-term profits that make the company look good to investors, and, finally, sell when the company can attract the highest price. That was the fate of Henry Disston & Sons.[51]

Within a few years after corporation-collector Thomas Mellon Evans took over Henry Disston & Sons, Tacony was irrevocably changed. Total employment was down to 1,100 from 1,822, with production workers numbering 542 instead of 1,500. Forty-two of the company's sixty-four acres were up for sale, and only thirty-two of the sixty-four buildings were being used. Sales were around $17.5 million, down from almost $24 million in the last two full years of the old management. Porter was happy with the company's financial situation, since Disston was "slightly" in the black. When this record is compared with Disston's loss of $5,000,000 in 1953 and $1,342,000 in the year before Porter's purchase, one can understand the new owner's satisfaction.[52]

Another dramatic change was in the workers' housing patterns. A factory complex that employed local residents almost exclusively (98 percent in 1900) had been transformed by post–World War II suburban expansion into a plant with few local workers. An H. K. Porter health insurance list that identified the 1972 address of each worker indicated that of 140 workers, only 25 still resided in Tacony, even though many of these workers had spent their lives at the plant.[53]

Evans had little feeling for former Disston employees. A petition was filed before U.S. District Court charging that their pension rights had been illegally terminated when Disston assets were sold to Porter. Thirty-five nonunion employees who said they had twenty-five years' service and were sixty-five years old before the sale date of November 16, 1955, demanded their pension options. A second suit was filed by sixty nonunion employees who were not yet age sixty-five but also had twenty-five years of service. The suit alleged that on September 1, 1950, Disston had adopted a pension plan for nonunion personnel. Charging that Evans's confiscation of the pension fund to pay off the PNB debt was "improper and illegal" and in violation of the employees' rights, the suits declared that the workers counted on the pension "as part of the consideration of their continued service." The plaintiffs stated that although Disston had sold all its physical assets it now had cash and securities valued at about $6.5 million and no liability other than the retirement benefits. They asked the court to force the company to honor the pension plan. The nonunion office workers won their claim but, unfortunately for the union workers and the community, the court ruled that H. K. Porter had a legal right to attach their pension funds because the employees had not contributed to them and there had been no specific agreement that the funds belonged to the union.[54]

Another suit awaited the Disston family. H. K. Porter filed suit to compel the directors of Henry Disston & Sons to adhere to the merger agreement of 1955, asking for $134,000 in damages from defendants Jacob S. Disston, William Leeds Disston, S. Horace Disston, Edwin J. Gillfillan, Richard T. Nalle, Henry Bain, John R. Wanamaker, Charles S. Redding, A. S. Corson, and Arthur F. Kroeger. These men were accused of failing to take steps to effect prompt dissolution of the corporation and distribution to its stockholders. On April 11, 1958, Common Pleas Judge Edward J. Griffiths ordered them to live up to the agreement with H. K. Porter. However, no fine was imposed on the Disstons, since they were to prove that the dissolution was legitimately delayed by unsettled accounts owed the Disston company.[55]

Evans decided in 1956–1957 to streamline his operation even further by consolidating divisions. Because he already owned a more profitable steel plant, Evans ordered Norman Bye to close the one at Disston. A special Real Estate Department was created and headed by Clyde Cassell, assisted by Roland Woehr, to sell off Disston property. Next, Nippes told Cassell that the lumberyard was too expensive to maintain and that saw handles could be made more cheaply by using less-expensive, precut furniture wood. The lumberyard was closed and sold.[56]

The plants in Toronto, Guilford, and Seattle continued in operation. A recommendation to enlarge the Toronto factory in 1947 had never been implemented, but business was brisk. Saws were important to the Canadian forestry industry; the same was true of the Seattle and Guilford operations. The latter two markets were particularly interested in the circular-saw business. Australian lumber publications often featured Disston employees, as in an article about master smither William Rowen in *The Australian.* These operations were closed or sold in the 1970s. Their modest success did not offset the financial burdens of the Tacony operation.[57]

In an effort to cut labor costs in saw production, Evans hired the Harvard-educated Paul Benke to survey the southern states to find a suitable location for a new, 1,500-square-foot factory. After visiting several sites, Benke settled on Danville, Virginia, because the town's Chamber of Commerce gave a favorable land purchase deal. The file, grass shears, and handsaw departments were moved to that location, and workers were asked to move or leave their job. This struck the Disston union as little more than union-busting, at a time when the union was fighting in court for pension rights for employees laid off from the steel and handsaw departments.[58]

In 1956 H. K. Porter sent Charles Norbeck to Danville. In his job in the Traffic Department he was responsible for the repair of defective hedge shears returned to the Tacony plant. Under the new corporate culture that seemed to believe it was cheaper to repair or replace goods than build them right the first time, Norbeck spent most of his time dealing with small errors made during the original production of the shears. Although he believed it would have been easier to build them right the first time, he doubted that Porter's new assembly line could do it. Eventually, Norbeck was moved to Maintenance when H. K. Porter moved the Traffic Department to Danville.[59]

The move created another problem for H. K. Porter. Machines from the Tacony plant were dismantled, reconditioned, painted, and then sent to Danville. New workers and college-trained engineers reassembled the machinery, only to find that they could not get it to work properly. The machines lacked the fine-tuning the skilled workers at the Tacony plant gave them, and more than half the items produced were useless. Finally, H. K. Porter had no choice but to bring to Danville as many original Disston/Tacony operators as possible to "get the machines to work right." The difficulty for H. K. Porter was that these were the very workers who had been let go by the company and denied their pensions. To sweeten the offer, Porter offered two months' overtime wages with free transportation and housing in Danville. To the union's consternation, five men accepted. They were able to get the machinery running properly by adjusting backward belts, adding proper weights for balance, and correctly setting gauges.[60] Returning to Tacony, they found themselves ostracized by the community. To this day some Tacony residents do not talk to one another because of the "Danville incident."

H. K. Porter based every business decision on short-term profits. As a matter of company policy, therefore, the searches for new locations were predicated on low labor costs. In June 1960 H. K. Porter, in an effort to find cheaper labor, opened a plant for handsaws in Colonia Mectezuma, Mexico. Equipment from the Philadelphia plant was sent to the new "H. K. Porter Co. de Mexico," where, during the first year of operation, approximately 15,000 No. 600 saws were manufactured. This early success led to an expansion of the product line that included saw handles and plastering trowels. Nevertheless, training remained a problem. The conditions in Mexico were stated clearly in a letter from Norman Bye to Albert Strasser, head of Porter/Disston Mexican operations:

> Once this mechanical work is done, the time for training operators will depend a great deal on the quality of the personnel available for training. . . . There are three major hand operations of straightening, assembly and blocking which require considerable time to learn and experience in working before they can be done properly. In Philadelphia, there was a regular two to three year apprentice course for trowel makers which covered all operations but, in Danville, the overall apprentice course was discontinued. There the hand operations alone still took from six to eight months before a man was considered fully qualified and could work entirely on his own or give instructions to others. It would not be expected that anything like this would be required in Mexico because of the lower labor cost per hour. If the work is done at a slow production rate, economically allowable in Mexico, satisfactory quality production can be achieved much more quickly. However, considerable time will still be required to permit satisfactory training to be performed.[61]

Early reports indicated that seven men and a foreman were able to produce 150 saws a day. Handle production included nonproductive techniques such as hand sanding, drilling holes individually, and finishing by spraying and sanding three times. Some improvements were being made in both machinery and developing jigs for drilling and slotting, but finishing was still a problem. Management was working on the problems, but "the solution seems a long way off."[62]

Trowel production added to the complexity of the training problem. Trowel handles still were not available seven months into the project. Equipment that was supposed to have been shipped from Danville and Philadelphia had not been received. Bye warned that if the initial 8,000 saw orders were a true indication of the Mexican production requirements, floor space would also be a problem for the new factory. All these factors eventually led to the failure of the Mexican plant.[63]

For the time being, however, with pruning operations completed and new factories functioning in Mexico and Danville, the company appeared to be financially stable. There were still the profitable old accounts—shipments to South America of chisel point circular saws and to the E. D. Barris Company in England of shears and handsaws. These customer sales and labor cutbacks, with the selling of unwanted equipment and property, allowed H. K. Porter to pay off the debts incurred in the

Disston purchase and to operate in the black. But the company had few skilled workers left and a skeleton crew in Tacony that was ill-prepared to meet future economic needs. The Tacony plant was, in a word, decimated.

At the half-deserted factory site, decay and ruin set in. Fires set in the unoccupied buildings were a serious threat to the community. On Thanksgiving Day 1962 a four-alarm fire destroyed a two-story, 250-foot-long building at 6801 Milner Street despite the efforts of 125 firefighters. It had been closed at the time of the fire.[64]

On October 12, 1970, a six-alarm fire burned three buildings on Unruh Street near the Delaware River. The fire was investigated because there had been two other multiple alarm fires in vacant buildings in a two-month period. Investigations indicated that the juveniles of the community were setting fires in many vacant buildings on the Disston site.[65]

About the same time at Tacony, H. K. Porter developed battery-operated grass-cutting shears that quickly catapulted from an unknown product to a multimillion-dollar business. Russell McIntyre recalls that between 1965 and 1968 the company could not produce enough shears to fill the orders. The quality of these shears gradually diminished, as did sales, but in 1970 this new product and a record of five straight years of profit made it the right moment, in Evans's judgment, to take Disston stock public. William L. Disston, then living in California, got a call from his stockbroker asking if he wanted to buy some newly offered Disston public stock. The company reports showed high profits, and the stock would sell for about $17 a share. Knowing the value to be inflated, William told his broker not to buy—considering the condition of the company, the stock was not worth more than $4 a share.[66]

The shares sold well from the start, and Porter quickly realized a profit of approximately $34 million. Evans stepped aside and handed the company over to the new president, John Poth. Under Poth, H. K. Porter fared less well. The weak infrastructure of the Disston operation soon began to show, and increased competition from less-expensive, foreign battery-operated shears cut into profits. The shoddy workmanship of the Disston shears further lessened sales. After the stockholders suffered great losses, Poth sold the Disston / Tacony plants to Sandvick, a Swedish saw firm, in 1978. The Danville plant was purchased locally by Virginia investors and now operates under the name Disston Company and is a national producer of saw blades, drills, and saws.[67]

The fate of the workers at Disston & Sons and the community was sealed when they became part of Thomas Mellon Evans's acquisition scheme. Although Evans's operations were legal, their morality was open to question. The English Victorian poet Robert Southey understood human nature and people like Evans when he wrote 150 years ago, "The greedy, grasping spirit of commercial and manufacturing ambition and avarice is the root of all our evils."[68]

Sandvick was less ruthless than H. K. Porter, but by this time there were few workers left at the Tacony plant and little to save. Nevertheless, the workers who remained, mostly those making circular saws, felt good about their new owners. For the first time since the Disston family sold the plant, the workers felt someone cared about them. There was a different attitude from the office down. Russell McIntyre remembers Sandvick's reconditioning the Marison grinder so it could do precision work. Whereas H. K. Porter did little to improve salaries, one of Sandvick's first priorities was to implement the "Hay Report," a management plan that recommended paying higher salaries for accomplishing goals. It had the effect of raising the salary of plant management first; only later, and never commensurately, did the workers' pay climb.[69]

In 1984 Sandvick sold what was left of Disston, Tacony, to R.A.F. Industries, led by Robert Fox of Jenkintown. The name Henry Disston & Sons was changed to Disston Precision Inc., a producer of specialty circular saws. Roland Woehr continued with the firm throughout these transactions. Woehr is currently chief engineer at the plant and in charge of production of specialized circular saws for cutting hot steel.[70]

The plant today consists of four buildings and twenty to thirty employees, depending on the orders to be filled. The specialty circular-saw business is the backbone of the operation, but specialty steel washers are a secondary product. Steel is cut by a torch and purchased according to specific order.[71]

What of the Tacony community? The stores along Longshore Street closed, with Torresdale Avenue businesses soon following suit. In the 1960s, gangs appeared on Torresdale Avenue for the first time, and by 1970 this once proud commercial center yielded to an assortment of secondhand furniture stores, boarded-up shops, and medical offices. Graffiti appeared on walls and windows along Longshore Street and Torresdale Avenue for the first time. Many changes in the community can be blamed: Interstate 95 split the community and permitted the younger generation to escape to the newly developing northeastern suburbs; new shopping malls drew customers away from corner stores; the homes of the community lack modern amenities, most having been built between 1880 and 1930; and television resulted in the closing of the Liberty movie house, thus decreasing traffic on Torresdale Avenue.

As important as any one of these factors was in promoting the demise of Tacony as an industrial town, the most influential factors in damaging the economic life of the community remained: the demise of Henry Disston & Sons and the subsequent denial of pensions to Disston retirees. This left the local population without money and with no faith in the future. Homes could not be maintained and the economic base necessary to sustain operations and retain customers was not present. Young couples avoided buying homes in the community. Home values were slow to keep pace with those in the nearby newer community of Mayfair. Older residents, many of whom were former Disston employees bereft of their pensions, were forced to remain in the area because they had only their social security benefits to live on. Many Taconyites were doomed to a meager existence in their declining years. It was not long before the community began to show the ravages of urban blight. All this occurred because profits outweighed human concerns at Disston when H. K. Porter controlled the company.

Epilogue

DESPITE THE FALL FROM GRACE of the saw industry in the post-industrial world, the shadow of Henry Disston still hovers over the Tacony community. Sandstone walls in every part of the community remind everyone, every day, of the town's debt to Disston. They are reminders of the beginning and end of an era of paternalistic management that created a way of life for workers and their families.

Paternalism could not endure untouched in an urban setting. The ingredients necessary for its success were too soon altered by larger populations, transportation improvements, and urban industrial expansion. Moreover, although a nineteenth-century proprietary capitalist could develop a paternalistic community, as Henry Disston did, the dilution of capital inherent in a one-family business left the company unable to maintain the paternalistic network in succeeding generations. Nevertheless, for three generations the idea endured.

In interview after interview, former Disston employees praised the Disston family for their benevolence and fairness. To this day, longtime Tacony residents have kind words for what the Disston saw works did for them. As for the deed restrictions, they are still in place and likely to remain so. A recent effort by Neil Rosenwald, son of founder Joseph, to gain a liquor license for Foodarama evoked a bitter reaction and protest. The arguments for family structure, community values, and a Christian life that were used to support the opposition to the license were similar to those of Henry Disston a hundred years before. The connection of Tacony to its past is anchored in this vision of what a good life should be.[1]

The age-old trade of saw-making has changed very little since Henry

Disston & Sons came to Tacony. Disston Precision Incorporated continues to make circular saws with the equipment left over from the turn of the century as the workers use the same skills. The skilled-labor shortage that Henry Disston faced in the 1870s remains an issue today.

Walter Arnold's smither ancestors have already been mentioned in this history. In 1973 Arnold retired, leaving no experienced smithers in the circular-saw department at H. K. Porter. The agreement between labor and management for shorter training periods had eliminated the once-productive apprenticeship system. The H. K. Porter company had no choice but to do what Henry Disston had done a hundred years before—turn to Sheffield, England, for skilled workers. In July 1973 it purchased an ad in the Sheffield *Telegraph & Star* offering jobs to skilled smithers. Disston would pay transportation and housing to any interested party. A ten-day trial period was to follow, and if service was satisfactory a job offer would be made.

Ron Turfitt, a thirty-five-year-old smither with twenty years' experience, read the ad, but since he had a good job at Spear-Jackson Saw Company in Sheffield, a home, and six children, he was not interested. His wife, Sheila, had other ideas. She thought America might offer the family greater opportunity, so she persuaded Ron to answer the ad and go to Tacony.

When Ron arrived in Tacony, he and another applicant, Dave Otter, were given a room and work. After ten days, both returned to England, leaving their telephone numbers with H. K. Porter representatives. Almost immediately, a call came to the Turfitts offering a loan of $1,800 to purchase a home in the 6500 block of Edmund Street, moving expenses, and transportation if Ron would take a full-time smithers job. Sheila got Ron to accept, and the family moved to Tacony.

Turfitt went to work the day after he arrived. The company had no one who could match his apprenticeship experience. Turfitt had entered his six-year apprenticeship when he was fifteen years old, working through the steel slump of the 1950s only to see the circular-saw business become strong again in the 1960s. He trained his son in the trade and had apprenticed many boys during his years at Spear-Jackson. He loved to train apprentices—not for the money, but just to "leave something behind. They can earn a living and there is a bit of myself in them. My apprentices have become like my sons."

Unfortunately, the Disston/Porter company did not see fit to hire apprentices to work with Turfitt, reasoning that it was too expensive. When asked what the firm will do for smithers in the future, he replied, "I honestly don't know."

Turfitt has had many "holders," but most eventually leave because of injury, such as the man whose arm was broken because the circular saw he was holding was struck when it was not sitting flat on the anvil. Smithing of circular saws has not changed in 150 years and is unlikely to change in the near future. The question that remains is where the skilled smithers will come from when men like Turfitt retire.[2]

The same dilemma was posed when the pocket-size "black books" used by skilled workers were discarded by H. K. Porter supervisors. It is clear that the traditional view of factory workers as using hands and not heads is inaccurate. Factory workers at Disston were much more thoughtful and less directed than writings on the subject suggest. There is a myth in America that depicts blue-collar workers as brainless tools of unions and management. Disston workers in the company's heyday determined quality, designed products, knew what the customer wanted, and had a say in the production of goods. Disston skilled workers opposed assembly-line production because it eroded the quality of their work. At times the Disston management system foreshadowed W. Edwards Deming's total quality movement, practiced in Japan during the 1950s. The long-unexplored question of quality in American industry needs another look by historians.

Yet another issue raised in the study is family ownership of American firms. The story of the Henry Disston company follows a typical trajectory. Family businesses had the disadvantage of fostering inbreeding and insularity and stifling creativity. With little capital for experimentation and a paucity of new ideas, such firms are vulnerable to any economic readjustment or downward turn. Dividend payments to family members become burdensome after three generations, at which point there are too many family members to be supported by the company. Jobs are invented, dividends are too high, and too many opinions compete, preventing a unified vision for the company. The seeds for the eventual sale of Henry Disston in 1955 were sown once the family shares were distributed.

Henry Disston created a way of life for Disston workers and the people of Tacony. So did the management of Henry Disston & Sons from afar by Thomas Mellon Evans. H. K. Porter stripped the company of the homebred qualities, ethics, and virtues that had made it successful from the days of Henry Disston through World War II. Evans's decision-making did not take into consideration customer needs or the value of meticulous quality-control in manufactured products. It took but a few years for what had been the mightiest saw works in the world to disintegrate. Destruction of the firm came from within, not from foreign shores.

At the root of the struggle between Evans and Tacony are the contradictions between the imperatives of capital and people's need for community and economic security. Planning theorist John Friedmann of the University of California at Los Angeles best describes this dilemma:

> Two geographies together constitute a "unity of opposites." I shall call this "life space" and "economic space." Although both are necessary for the sustenance of modern societies, they are inherently in conflict with each other. Over the last two centuries, economic space has been subverting, invading, and fragmenting the life spaces of individuals and communities.
>
> Life space is at once the theater of life, understood as a convivial life, and an expression of it. . . . Life spaces exist at different scales [and] are typically bounded, territorial spaces. . . . Places have names. They constitute political communities.
>
> In contrast, economic space is abstract and discontinuous, consisting primarily of locations (nodes) and linkages (flows of commodities, capital, labor, and information). As an abstract space, it undergoes continuous change and transformation.[3]

The story of Henry Disston & Sons offers yet another window through which to view the life-space versus economic-space dilemma. The life-space is the Tacony community, a fixed and interactive social arrangement of houses, schools, institutions, and a way of life bound to and generated by an industrial complex. It was sustained for more than eighty years by the mutual bonds between that industry and its people. Steeped in the belief that what was good for the worker was good for the industry, Henry Disston provided the community with both an economic base and a wholesome way of life. His humane concern for individual employees, exemplified by the hiring of those who could not get work elsewhere and his provisions for housing the workforce, were an expression of his belief in the value of life-space to industry. Although his motives were related to the shortage of labor in the community, his policies promoted a dignified, healthy, and productive life-space. Commitment to the community was maintained by the Disston family as late as the depression of the 1930s, when short workweeks divided available wages equally among the firm's most valued, long-term workers. The family's dilemma in the 1950s centered on their inability to project expansive and innovative products for the future and still provide the proper life-space for the workers.

Thomas Mellon Evans cared little about the issue of life-space. As a businessman in a distant city, he was only concerned about economics. He viewed the purchase of Henry Disston & Sons saw works as a means to make money, and his plan depended on a view of flexible and impersonal economic space.

The legacy that Disston & Sons leaves the world is in blueprint form. It is a model that beckons future generations to find a way to convince future leaders of industry, even those from high-tech knowledge-based firms, of the value of their most important resource—the community around them and the people who produce and consume their products. To do this, a consideration of life-space is imperative. The story of Henry Disston & Sons is living proof of what both benign and ruthless capitalism can do to a community.

Appendix A

Disston Company Name History

1840 **Keystone Saw Works, Philadelphia**

1850 First catalog title:

Keystone Saw Works, Philadelphia
Henry Disston

1865 First son, Hamilton Disston, had served seven years as an apprentice beginning in 1858:

Keystone Saw Works, Philadelphia
Henry Disston & Son

1871 In 1871 Albert Disston, who had been in the accounting department, was given responsibility for the financial management of the company:

Keystone Saw Works, Philadelphia
Henry Disston & Sons

1886 Sons Horace C., William, and Jacob all joined the firm after serving apprenticeships, and the business was incorporated:

Keystone Saw Works, Philadelphia
Henry Disston & Sons Inc.

1955 Henry Disston & Sons Inc. sold to H. K. Porter in Pittsburgh and named:

Carlson Rules and Measures / Henry Disston
Division of H. K. Porter Inc.

1978 Carlson Rules and Measures / Henry Disston Division of H. K. Porter Inc. sold and renamed:

Henry Disston Division of Sandvick Saw of Sweden

1984 Sandvick Saw sells Henry Disston Division to R.A.F. Industries, owned by Robert Fox and named:

Disston Precision Incorporated

Appendix B

Disston Company Presidents

Henry Disston (founder)	1840–1878
Hamilton Disston (first son)	1878–1896
William Disston (fourth son)	1896–1915
Frank Disston (son of Albert)	1915–1929
Henry Disston (grandson, son of Hamilton)	1929–1938
*William D. Disston (grandson, son of William)	1931–1937
S. Horace Disston (nephew, son of Samuel)	1938–1948
Jacob Disston Jr. (grandson, son of Jacob)	1948–1954
**John D. Thompson	1954–1955
William Leeds Disston (son of William D.)	left firm 1955†
Henry Disston III (son of Jacob Jr.)	left firm 1956†

SOURCE: The above list of Disston presidents was given to me by William L. Disston. He used the "Disston History" and his own memory to reconstruct the changes in leadership.

*When Henry Disston suffered a stroke in 1931, William D. Disston became acting president at the factory site. When William D. Disston had a heart attack in 1937, S. Horace Disston filled in and in 1938 became president.

**The only non-Disston family man to head the firm.

†The split in the family at the time the company was sold occurred between the descendants of Jacob and William Disston. It is interesting, however, that two of the most helpful Disstons were Morris Disston (son of William L. Disston) and Henry Disston IV (son of Henry Disston III and grandson of Jacob Jr.). Despite all that had gone before, they were the best of friends and wonderfully supportive.

NOTES

Chapter 1

1. There is no biography of Henry Disston. William Dunlap Disston, Henry W. Disston, and William Smith wrote "The Disston History" (vol. 1) and "The Disston History: The Family" (vol. 2), which were compiled by Elizabeth B. Satterthwaite in 1920 and are in the Eleutherian Mills Historical Library, Hagley Museum, Greenwood, Delaware. These two volumes are hereafter referred to as "Disston History," followed by the volume number. For the section about Disston learning his trade from his father, see vol. 1, 34. For sketches of Disston's life, see "The Keystone Saw, Tool, Steel, and File Works," in *Annual Report of the Secretary of Internal Affairs of the Commonwealth of Pennsylvania, Part III: Industrial Statistics,* vol. 15 (Harrisburg, Pa., 1887), E26–29; Jacob S. Disston Jr., *Henry Disston, 1819–1878: Pioneer Industrialist, Inventor, and Good Citizen* (Philadelphia: Author, 1950); John T. Scharf and Thompson Westcott, *History of Philadelphia, 1609–1884* (Philadelphia, 1884), 2267–68.

2. *Philadelphia Public Ledger,* March 18, 1878; *Philadelphia and Popular Philadelphians* (Philadelphia, 1891), 133; Scharf and Westcott, *History of Philadelphia,* 2267–68; Philip Scranton and Walter Licht, *Work Sights: Industrial Philadelphia, 1890–1950* (Philadelphia, 1986), 170–87; *Wood-Worker* 9 (December 1890).

3. "Disston History," 1:35–37; *Wood-Worker* 9 (December 1890).

4. Paul W. Morgan, "The Henry Disston Family Enterprise I," *Chronicle of the Early American Industries Association* 38 (June 1985), 18. Stories of this kind were commonly told about successful people in Disston's day to illustrate the maxim that hard work, honesty, and virtue would be crowned with success.

5. Geoffrey Tweedale, *Sheffield Steel and America: A Century of Commercial and Technical Interdependence, 1830–1930* (New York: Cambridge University Press, 1987), 25.

6. Edwin T. Freedley, *Philadelphia and Its Manufactures: A Handbook Exhibiting the Development, Variety, and Statistics of the Manufacturing Industry of Philadelphia* (Philadelphia, 1859), 330. See also "United States Circuit Court for the Eastern District of Pennsylvania, April 1859," Equity Case 10, *Disston v. Cresson* (National Archives, Mid-Atlantic Branch, Philadelphia).

7. *Philadelphia Press,* January 15, 1861. See also "U.S. Circuit Court for the Eastern District of Pennsylvania, April 1859," Equity Case 10, *Disston v. Cresson;* Morgan, "Disston Family Enterprise I," 18–19.

8. *The Disston Crucible: A Magazine for the Millman,* April 1919, 42–43. (Copies of the *Crucible* are available at the Philadelphia Free Library, Logan Square.)

9. J. Matthew Gallman, *Mastering Wartime: A Social History of Philadelphia During the Civil War* (New York: Cambridge University Press, 1990), 266–67; Nicholas B. Wainwright, *A Philadelphia Perspective: The Diary of Sydney George Fisher* (Philadelphia:

Historical Society of Pennsylvania, 1967), 88–90, entries for May 7, 11, and 19, 1861. See also Allen Nevins, "A Major Result of the Civil War," *Civil War History* 5 (September 1959), where Nevins establishes two themes: that the North's economy was not ready for the war, and that the war would eventually help Americans adapt to large organizations. This knowledge of how to manage large institutions would lead to centralization of government and business when the war ended.

10. Wainwright, *Philadelphia Perspective,* 442, entry for November 26, 1862; Gallman, *Mastering Wartime,* 1990, 271.

11. *Crucible,* April 1919, 4; Gallman, *Mastering Wartime,* esp. chap. 11, "The Economic Life of Wartime Philadelphia," 266–98.

12. Glenn Porter and Harold Livesay, *Merchants and Manufacturers: Studies in the Changing Structure of Nineteenth-Century Marketing* (Baltimore: Elephant Paperbacks, 1971), 119–21.

13. Scharf and Westcott, *History of Philadelphia,* 2338–39; Gallman, *Mastering Wartime,* 281.

14. "Disston History," 1:17–19, 30; Morgan, "Family Enterprise I," 18–19; *Disston Lumberman's Handbook,* rev. ed. (Philadelphia, 1923), 3–12 (originally published in 1870).

15. Tweedale, *Sheffield Steel,* 147–48. Tweedale warns that "technological sophistication of the American saw industry must not be exaggerated." Skilled labor was always a necessity in saw-making. Smithing, for instance, never became mechanized, and skilled workers were always part of the workforce at Disston.

16. Lorin Blodget, *Manufactures of Philadelphia, Census of 1860* (Philadelphia, 1861); Lorin Blodget, *Manufacturers of the City of Philadelphia, Census of 1870* (Philadelphia, 1877); Edwin T. Freedley, *Philadelphia and Its Manufactures* (Philadelphia, 1867), 581–84; Gallman, *Mastering Wartime,* 258–59, 284–85.

17. "Disston History," 2:35, lists the children of Henry and Mary Steelman Disston: Hamilton Disston, August 23, 1844–April 30, 1896; Amanda Disston, July 13, 1846–June 19, 1851; Franklin Disston, August 26, 1847–November 8, 1848; Albert H. Disston, July 23, 1849–October 21, 1883; Amanda F. Disston, July 22, 1851–December 6, 1851; Mary Disston, November 10, 1853–April 16, 1878; Horace C. Disston, January 10, 1855–June 13, 1900; William Disston, June 24, 1859–April 15, 1915; Jacob S. Disston, August 4, 1862–February 25, 1938. (Not all children had middle names.)

18. "Disston History," 2:1–40.

19. Morgan, "Family Enterprise II," *Chronicle of the Early American Association* 23 (September 1983), 42–44; Morgan, "Family Enterprise III," in ibid., 23 (December 1985), 68; J. S. Disston Jr., *Henry Disston,* 8. Jacob Disston writes proudly of Henry's English heritage: "In a sense, we can recognize in the story of this man Henry Disston's career the linking of England and the United States in a common bond. He well may symbolize for us a life that continued in this nation the heritage of British workmanship, character, fair play and human tolerance" ("Disston History," 1:68–71). These partnerships were part of Henry Disston's pattern of familial benevolence toward his brothers and wife. As each of the men died, the partnership was bought out by Henry Disston & Sons, with the final purchase occurring in 1899.

20. J. S. Disston Jr., *Henry Disston,* 19; "Disston History," 1:21–22, 39.

21. Robert Van Dervort, *Tacony* (Philadelphia: Free Library, 1982); Freedley, *Philadelphia and Its Manufactures,* 330; John T. Faris, *Old Roads Out of Philadelphia* (Philadelphia: Free Library, 1917), 285–302; Michael Feldberg, *The Turbulent Era: Riot and Disorder in Jacksonian America* (New York: Oxford University Press, 1980), 65–68.

22. Francis X. Roth, *History of St. Vincent's Orphan Asylum, Tacony, Philadelphia: A Memoir of Its Diamond Jubilee, 1855–1933* (Philadelphia: Nord-Amerika Press, 1934), 15–19. There is no mention in the family history of Thomas owning land in Tacony, but his name does appear on the original land records of St. Vincent's and in the censuses of 1870 and 1880.

23. "Disston History," 1:52–54; Samuel L. Smedley, *Atlas of the City of Philadelphia . . . 1862*, on 23rd Ward (Tacony area); Morgan, "Family Enterprise II," 43–45; U.S. Census, 1880, Philadelphia, 35th Ward (National Archives); George Smedley Webster, "Plan of Building Lots Belonging to Henry Disston and Sons, Philadelphia 1890," Disston Collection, Eleutherian Mills Historical Library; R. A. Smith, *Philadelphia as It Is in 1852, Being a Correct Guide to All the Public Buildings, Literary, Scientific, and Benevolent Institutions, and Places of Amusement* (Philadelphia, 1852), 411. In describing the Philadelphia & Trenton Railroad, Smith writes: "Passengers by this road leave by foot of Walnut Street steamboat, for Tacony, thence by railroad through Bristol and Morristown to Trenton, thence by Trenton and New York Railroad . . . to New York."

24. *Annual Report of the Secretary of Internal Affairs*, E25–E30; Morgan, "Family Enterprise I," 19.

25. J. S. Disston Jr., *Henry Disston*, 16–17.

26. Interview with Catherine Seed, April 14, 1989; *Tacony Souvenir Program, Prepared for the Grand Celebration at Tacony Week of May 30th, 1906* (Philadelphia: New Era Press, 1906); U.S. Census, 1880, Philadelphia, 35th Ward.

27. U.S. Census, 1880, 35th Ward, Philadelphia Free Library, Logan Square. There were few streets in Tacony at the time of the census. For an example of the importance of English customs in the community, see the picture of the Washington Tea House (Figure 31).

28. Tweedale, *Sheffield Steel*, 150, 175.

29. Right away Hamilton Disston showed some of the sales ability of his father. One month after Henry Disston's death, he entertained President Rutherford B. Hayes, an event well covered and discussed throughout the city. "Disston History," 1:4–5; *Crucible*, June 1914, 74–80. Obituary of Henry Disston, *Philadelphia Press*, March 17, 1878.

30. "Disston History," 1:18–40. See also John A. Fitch, *The Steel Workers* (Pittsburgh: University of Pittsburgh Press 1989), 130–32.

31. Chris McGuffie, *Working in Metal Management and Labor in the Metal Industries of Europe and the U.S.A., 1890–1914* (London: Merlin Press, 1986), esp. chap. 2.

32. David A. Hounshell, *From the American System to Mass Production, 1800–1932: The Development of Manufacturing Technology in the United States* (Baltimore: Johns Hopkins University Press, 1984), esp. chaps. 4–8.

33. Federal Writers Project of the Works Progress Administration (WPA) for the Commonwealth of Pennsylvania, *A Guide to the Nation's Birthplace* (Philadelphia: Natural Living Press, 1937), 528–30.

34. Scranton and Licht, *Work Sights*, 170–81; interview with Russell McIntyre, December 27, 1990. McIntyre says that the work in the filing department changed little over the seventy-five years it operated. Disston did not buy new equipment, and the daily job routines of the workers were handed down from one generation to another. This, he contends, was part of Disston's eventual financial crisis.

35. Interview with Russell McIntyre, December 27, 1990.

36. "Disston History," 4–5; *Crucible*, June 1914, 76–79.

37. Henry Disston & Sons fire map, 1944, Atwater Kent Museum.

38. J. S. Disston Jr., *Henry Disston*, 17; "Disston History," 1:72–73.

Chapter 2

1. "Disston History," 1:110.

2. Henry Disston to "Fellow Workers," November 13, 1867, "Disston History," 1:32. The letter was written to thank the workers for their gift of silverware.

3. *Disston "Bits": A Monthly Magazine of "Inside Stuff,"* September 9, 1919, published by Henry Disston & Sons. The *Disston "Bits"* was a company-sponsored monthly newspaper for workers. Bound copies covering the years 1917–1919 are at the Atwater Kent Museum, Philadelphia.

4. *Disston "Bits,"* March 1918, 85; *Wood-Worker* 9 (December 1890), 25. David Bickley, Disston's first worker, stayed on at Keystone Saw Works for more than forty years, rising to the position of foreman of the long-saw department. In 1890 he told how Disston toiled side by side with him in the 1840s and 1850s, sharing work equally.

5. "Disston History," 1:5–6, 56–57, 82–85; *Public Ledger,* March 18, 1878.

6. "Disston History," 2:35.

7. Disston Real Estate Sales Book No. 1, in the possession of Louis A. Iatarola.

8. One observer commented, "In the spring of the year, when flowers are in bloom, the effect must be both pleasing and striking. It attracts the attention of passengers on the numerous trains that pass Tacony." Such parks were the "lungs of the city," the author later stated, adding that every village should begin by laying out a park, "for parks give those around them the advantage of light and air." "History of Olde Tacony" (1920), reprinted in S. F. Hotchkin *The Bristol Pike* (Philadelphia: Fidelity Federal Savings & Loan, 1950), 4–5; booklet found in the Tacony Free Public Library.

9. Interview with Louis A. Iatarola, April 13, 1990. In 1990 Iatarola restored the Music Hall to its 1885 status. Located on Longshore Street east of Torresdale Avenue, the hall remains today the center of "old" Tacony.

10. "Disston History," 1:70–81. For information about Tacony, see *Crucible,* June 1913 and August and September 1915.

11. Van Dervort, *Tacony,* 11; "Disston History," 1:42–43. See also "Social Base Map—Philadelphia," 1939, WPA, Urban Archives, Temple University. In 1939 there were still no saloons on the land in Tacony that was part of Disston deed-restricted land. *Philadelphia Inquirer,* January 21, 1988, "Neighbors/Northeast Philadelphia" section, 1–3. In 1988 Neil Rosenwald, a local caterer, was refused a liquor license because of the Disston deed.

12. "Disston History," 1:107.

13. Ibid., 45.

14. Ibid., 32.

15. This family contact, and the control of the company by descendants of Henry Disston for 150 years, kept remnants of the paternalistic system alive until the 1940s. Family leadership over the years is shown in Appendix B.

16. Interview with Henry Disston (grandson of Jacob Jr.), July 12, 1988. Because church bells alerted small communities to fires, Henry did not want to repeat the Laurel Street experience of having men, including his own son, leave work.

17. Van Dervort, *Tacony,* has the deed. See also "Social Base Map—Philadelphia" (WPA, 1939), Urban Archives, Temple University. In 1939 there were still no saloons on the land in Tacony that was part of Disston deed-restricted land, and in 1988 Neil Rosenwald, a local caterer, was refused a liquor license because of the Disston deed (*Philadelphia Inquirer,* January 21, 1988, "Neighbors/Northeast Philadelphia" section, 1–3).

18. The sons of Lawrence Donohue and Jake Hepp worked at Disston in 1900 as apprentices, with no contract. They noted that throughout the history of the firm the most trusted workers were those who began as apprentices (interview with Larry Hepp, August

16, 1988, and Lawrence "Ricky" Donohue, July 1, 1989). For articles about these families, see *Disston "Bits,"* July 1919, 6; May 1920, 168; and "Disston History," 1:13, 32, 40, 50, 64.

19. Interview with William Rowan, December 7, 1988. See also "Disston History," 1:30–32, 42–45, 62–63; *Public Ledger,* March 18, 1878, 56.

20. David Roberts, *Paternalism in Early Victorian England* (New Brunswick, N.J.: Rutgers University Press, 1977), 4–5. Roberts describes paternalism as "an outlook held by landowners, captains of industry, clergymen, members of Parliament, justices of the peace, civil servants, newspaper editors, novelists, poets, and university dons. It was even held, as habits of deference, by agricultural laborers, operatives, and the worthy poor. It informed social attitudes at all levels of society and expressed itself in countless ways."

21. Quotations in the text are from Robert Southly's *Sir Thomas More; or, Colloquies on the Progress of Society* (London, 1832), which contained many pleas for a more paternalistic society; Samuel Taylor Coleridge's *Two Lay Sermons* (1816; reprint, London, 1839), which explored the philosophical and moral basis for paternalism; and Thomas Arnold's *Letters on Our Social Condition* (London, 1832), which called for reform of the church by joining churchmen and earnest manufacturers in a paternalist crusade to help the poor. Books on paternalism in America include Tamara K. Hareven and Randolph Langenbach, *Amoskeag: Life and Work in an American Factory-City* (New York: Pantheon Press, 1978); Thomas Bender, *Toward an Urban Vision: Ideas and Institutions in Nineteenth-Century America* (Baltimore: Johns Hopkins University Press, 1976); and Keith C. Petersen, *Company Town: Potlatch, Idaho, and the Potlatch Lumber Company* (Pullman: Washington State University Press, 1987). For yet another cultural view of paternalism, see John Bennett and Iwao Ishino, *Paternalism in the Japanese Economy: Anthropological Studies of Obabun-Kobun Patterns* (Minneapolis: Greenwood Press, 1963). Contemporary definitions of paternalism tend to be negative. Defined as a kind of "benevolent despotism," paternalism has become a symbol of an outdated practice in a more global and less autocratic corporate business world. Nevertheless, paternalism, which has its origins in religious beliefs in a "benevolent father" who looks after the well-being of his children and subjects, was originally viewed as a positive force. The strength of the church in Europe before the twentieth century also helped promote paternalism, urging people to value their stewardship over fellow human beings, first in religion, then as land owners, and finally as businessmen.

22. Ken Fones-Wolf, *Trade Union Gospel: Christianity and Labor in Industrial Philadelphia, 1865–1915* (Philadelphia: Temple University Press, 1989) 42, 106.

23. "Disston Picture Album," envelope containing newspaper clippings, now in the possession of William L. Disston (see article about Atlantic City).

24. "Disston History," 1:30–32, 42–45, 62–68.

25. Ibid., 4–5. Theodore Shoemaker, superintendent of the Disston works in 1888, wrote the section in "Disston History" on the strikes. He does not mention the 1884 dispute but does refer to the strikes of 1877 and 1886.

26. Ibid.

27. Hounshell, *From the American System to Mass Production,* esp. chaps. 4–8; Scranton and Licht, *Work Sights,* 170–81. Hareven and Langenbach, *Amoskeag;* Henry David, *History of the Haymarket Affair: A Study in the American Social-Revolutionary and Labor Movements* (New York: Collier Press, 1964); Paul Aurich, *The Haymarket Tragedy* (Princeton: Princeton University Press, 1984); Frederic Trautmann, *The Voice of Terror: A Biography of Johann Most* (Westport, Conn.: Greenwood Press, 1980), 123–39. Trautmann claims: "The events of 1885–1886 gave an opportunity to shout *anarchist.* Two million unemployed, cold and hungry, spread idle hands before the fires of radicalism." The

radical Philadelphia newspaper *Frei Wacht* defended those involved in the Chicago bombing, advocating labor unrest in the city until workers' rights were recognized.

28. *Taggart's Sunday Times* (Philadelphia), May 4, 1884. The story of the firing of Broomhead and the hiring of Williams is told to the public in this edition.

29. Ibid.

30. Fitch, *The Steel Workers,* 130–32.

31. Philip Scranton, "Varieties of Paternalism: Industrial Structures and the Social Relations of Production in American Textiles," *American Quarterly* 36 (Summer 1984), 235.

32. *Disston "Bits,"* March 1921.

33. Norman J. Ware, *The Labor Movement in the United States, 1860–1895: A Study in Democracy* (Gloucester, Mass.: Peter Smith Press, 1929), Introduction, xi–xviii; "Disston History," 1:15; Scranton, "Varieties of Paternalism," 235.

34. Stanley Buder, *Pullman: An Experiment in Industrial Order and Community Planning, 1880–1930* (New York: Oxford University Press, 1967), 38–59.

35. Ibid., 43.

36. Ibid., 216.

37. Ibid., 217.

38. "The Keystone Saw, Tool, Steel, and File Works," in *Annual Report of the Secretary of Internal Affairs* (see Chapter 1, note 1), vol. 15, E35.

39. Ibid., E25.

40. Ibid., E26, E27, E35.

41. Interview with Catherine Seed, August 24, 1989.

42. Smedly, *Atlas of the City of Philadelphia . . . 1862* (Philadelphia, 1862); G. M. Hopkins & Co., *City Atlas of Philadelphia . . . 1876,* vol. 3 (Philadelphia, 1876); G. M. Bromley & Co., *Atlas of Philadelphia . . . 1894,* (Philadelphia, 1894), vol. 13, 23rd and 35th Wards. It is difficult to calculate how much money was actually realized by the Disston family from their Tacony real estate. However, there is a record book of real-estate sales by Henry Disston's estate from 1872 to 1895 (Disston Real Estate Sales Book No. 1), which lists every property sold by the Disston family in Tacony. Total sales for that period were $701,368.89. Much land was left to Mary Disston and the family, all of which was sold after 1895. The best guess is that the development of real estate in Tacony netted the Disston family well over $10 million by the final sale of property in 1943.

43. Buder, *Pullman,* 158–59.

44. Ibid., 144, 158.

45. Ibid., 100–145; interview with Catherine Seed, August 24, 1989.

46. Buder, *Pullman,* 44.

47. "Disston History," 1:95–100; Morgan, "Family Enterprise III," 23, 77–79.

48. *Tacony Souvenir Program. . . . Tacony Week of May 30th, 1906* (Philadelphia, 1906), 6–9. The program cost 5 cents and was printed at the *New Era* newspaper office located at State Road and Knorr Street.

Chapter 3

1. Disston Private Ledger No. 2, 72, Disston Precision Inc. files; Real Estate Sales Book No. 1, 216–18, presently in the possession of Louis A. Iatarola; Henry Disston's Atlantic City Real Estate Book, Disston Collection, Atwater Kent Museum. The Real Estate Sales Book No. 1 contains a listing of all monies spent to develop Disston's Tacony real estate from 1872 to 1894.

2. Disston Letter Book, 362, July 3–August 31, 1872, Atwater Kent Museum. See also Disston Private Ledger No. 3, 185, Disston Precision files.

3. Atlantic City Real Estate Book, 3. This book also contains all purchases of equipment for the Disston company from 1867 to 1868.

4. Disston Private Ledger No. 2, 72–80, Disston Precision files.

5. Holmesburg Library Scrapbook, Holmesburg Public Library, Philadelphia; Disston Private Ledger No. 3, 79. Descriptions of each building and its use are found in the Heximer Maps, Philadelphia Free Library, Logan Square.

6. Minutes Book of the Tacony Fuel Gas Company, Atwater Kent Museum, Philadelphia.

7. Ibid., 16.

8. Henry Disston & Sons to Tacony Fuel Gas Co. board of directors, June 8, 1897; John Roberts to William Miller, June 25, 1897, found in Minutes Book of Tacony Fuel Gas Company, 24–25.

9. Interview with Elenore Shuman Dick, January 14, 1989.

10. Interview with Bob Bachman, September 18, 1988; Edward Darreff, August 27, 1988; and Harriet Seed, September 24, 1989.

11. William Lerner, *Historical Statistics of the United States, Colonial Times to 1970, Part 1* (Washington, D.C.: Bureau of the Census, 1975), 165.

12. Disston's management style can be determined by comparing what Disston did with the recent ideas of Tom Peters. See Tom Peters and Bob Waterman, *In Search of Excellence: Lessons from America's Best Run Companies* (New York: Harper & Row, 1982).

13. Disston Letter Book, July 3 to August 31, 1872; Atlantic City Real Estate Book; Private Account Books, 1867–1955, Disston Precision files. For the most comprehensive study of business practices in American business, see Alfred Chandler, *The Visible Hand: The Managerial History of America* (Cambridge, Mass.: Harvard University Press, 1977).

14. Interview with William Disston, May 13, 1990; Letters written by Disston, in Disston Letter Book, 54, 63, 64, 72, 82, 101, 125, 157, 158, 159, 195, 224, 230, 251, 253, 273, 307, 325, 351. Henry Disston was not a prolific writer. His sentences rarely had periods, and his use of capital letters was sporadic. Most letters began with "Respected Sir."

15. Morgan, "Family Enterprise I," 18.

16. "Henry Disston & Co. (Phila.) Patent & Design Drawings for Saws," Disston Collection, Acc. 1675, Eleutherian Mills Historical Library, Hagley Museum, Greenwood, Delaware.

17. Disston Letter Book, 87, 135, 204; *Crucible,* June 1913–June 1923.

18. Disston Letter Book, 63, 72, 82, 125, 253.

19. Ibid., 92, 136, 356.

20. Ibid., 356, 444. For examples of letters concerning saw repair, see ibid., 52, 57, 61, 72, 79, 81, 87, 100, 135, 149, 180, 342, 385.

21. Ibid., 82.

22. Ibid., 44.

23. Ibid., 174, 271.

24. Ibid., 60.

25. Ibid., 65, 66.

26. *Disston Lumberman's Handbook* (1923 ed.), 17–141; Disston Letter Book, 19, 25, 34, 51, 76, 77, 92, 204.

27. Disston Letter Book, 51, 81, 126, 135, 204, 205.

28. Ibid., 274.

29. Ibid., 72.

30. Ibid., 86.

31. Ibid., 69.

32. Ibid., 51, 69.

33. Ibid., 34, 40, 52, 65, 67.
34. Ibid., 230.
35. Ibid., 52, 75.
36. Ibid., 82; Disston Private Ledger No. 2, 72–80.
37. Ann (Harrod) Disston remarried after the death of her husband Thomas in Philadelphia. Her second husband was Samuel Newcome, who fathered Samuel. Henry Disston and the rest of the family treated Sam like a brother. Samuel's influence on the family was important. He was the father of S. Horace Disston, who became president in 1938. After Samuel's first wife, Sara, died, he married Jennie Cherry on April 29, 1874. Frederic H. Godcharles, *Encyclopedia of Pennsylvania Biography* (New York: Lewis Historical Pub. Co., 1932), 7:2581–82.
38. Disston Letter Book, 42, 206, 217, 218, 220, 221, 262; "Disston History," 1:80–84; W. A. Linas to S. Horace Disston, July 30, 1941. Letter found in "Record of Activities Dedicating New Disston Central Power Plant and New Philadelphia U.S.A. Armor Plant," June 16, 1941, Atwater Kent Museum.
39. See J. S. Disston Jr., *Henry Disston.*
40. "Disston History," 1:43–57; Disston Private Ledger No. 2.
41. Disston Letter Book, 159, 253.
42. Ibid., 411.
43. *Crucible,* May 1915, 53; *Public Ledger,* February 22, 1915; Disston Letter Book, 195–200.
44. Disston Letter Book, 101.
45. Quotations from the Disston Private Ledger No. 2, and Disston Letter Book, 237, 252.
46. Disston Private Ledger No. 2.
47. "Disston History," 1:95–111. Pictured are the company's fifteen branch offices in Chicago, Boston, Louisville, Bangor, New Orleans, Memphis, Cincinnati, Toronto, New York, Seattle, Sydney (Australia), Spokane, San Francisco, Vancouver, and Portland.
48. "Disston History," 1:110–12. Celebrations usually included parades. The following are examples of the parades that Disston & Sons took part in: a July 3, 1876, twilight "Celebration of the 100th Anniversary of Independence," a parade organized to celebrate the return of Ex-President Ulysses S. Grant from his trip around the world—December 16, 1879; a bicentennial celebration on October 25, 1882, a celebration of the 200th anniversary of the landing of William Penn in Philadelphia; and the celebration of peace at the close of the Spanish-American War, October 26, 1898.
49. *Philadelphia Made Hardware* (1913). (Presently in the possession of William L. Disston, this advertising catalog was published for Henry Disston & Sons, Miller Lock Co., North Brothers Manufacturing Company, Fayette R. Plumb Inc., and Enterprise Manufacturing Co. of Penna.)
50. Advertisement for "P.-M. H. Flying Window Trim" in *Philadelphia Made Hardware,* vols. 13–14 (October 1916–September 1917), 6. The history of advertising has been the subject of few studies, but Roland Marchand has studied advertising and its effects on the consumer (*Advertising the American Dream* [Berkeley and Los Angeles: University of California Press, 1985]). The *Philadelphia Made Hardware* and the *Crucible* point to the need for a study of advertising as it related to retailers. National brand names emerged from this process, allowing certain companies to capture entire product lines.
51. Mary Demont, "A Woman's Point of View," *Philadelphia Made Hardware,* vols. 13–14 (October 1916–September 1917), 6.
52. "Rest Rooms for Women Increase Business," in ibid., 9.

53. George M. Rittelmeyer, "Sparks from the Anvil" and "Good Advertising Contest Award," in ibid., 9.

54. *Crucible,* July 1913, 96; November 1914, 158–59; November 1915, 154–59.

55. Tweedale, *Sheffield Steel,* 152.

Chapter 4

1. *Philadelphia Record,* May 1, 4, and 5, 1896; Charles Horner, *Florida's Promoters* (Tampa, Fla., 1973), 12–15. *Frank Leslie's Illustrated Newspaper,* May 18, 1878; *Crucible,* April 1919, 42–43.

2. Horner, *Florida's Promoters,* 12–16.

3. Ibid., 15; Morgan, "Family Enterprise II," *Chronicle* (September 1985), 43–45.

4. Horner, *Florida's Promoters,* 14; interview with Henry Disston, grandson of Jacob Disston Jr., July 12, 1988.

5. *Philadelphia Record,* May 1 and 4, 1896; interview with Henry Disston, July 12, 1988; Horner, *Florida's Promoters,* 15.

6. "Disston History," 1:82.

7. Ibid., 80–84.

8. Interview with William L. Disston, October 23, 1990.

9. Disston Private Ledger No. 2, 72–80; interview with Roland Woehr of Disston Precision Inc., August 17, 1989.

10. "Disston History," 1:71.

11. Disston Private Ledger No. 4 (1898), 112, and No. 2 (1884), 212, Disston Precision files.

12. Journal No. 1, H. Disston & Sons, 1909, Disston Precision files.

13. *The Financiers of Philadelphia: A Practical Directory of Directors* (Philadelphia, 1900), 46; *Time,* May 27, 1940.

14. Disston Private Ledger No. 1 (1867–1869), 23, and No. 3, 79–80.

15. *Tacony Souvenir Program* (see Chapter 2, note 48).

16. *Tacony New Era,* April 14, 1887, presently in the possession of Louis Iatarola.

17. Samuel C. Willits, "History of Lower Dublin Academy," 1:133–34, Holmesburg Public Library, Philadelphia; J. L. Smith, *Atlas of the 23rd, 35th, and 41st Wards of the City of Philadelphia* (Philadelphia, 1910). The best single summary of Philadelphia manufacturing is *Workshop of the World: A Selective Guide to the Industrial Archeology of Philadelphia* by the Oliver Evans Chapter of the Society for Industrial Archeology (Philadelphia, January 1990); chap. 8, by Harry C. Silcox, contains a description of Tacony factories.

18. "When 1914 'Fire' Became Real, Did Hollywood-on-Schuylkill Die," *Philadelphia Bulletin,* April 30, 1967; "Aluminum Electroplating in Architecture," *Scientific American,* October 22, 1892, 261; "Tacony Iron Company," in *Tacony Souvenir Program.* Kern Dodge built a steel plant on the same site in 1919; see *Philadelphia Bulletin,* "They Know Casting," November 5, 1952.

19. "Erben-Harding Co.," in *Tacony Souvenir Program.*

20. "The Manufacture of Wire Glass at Tacony," *Scientific American,* November 5, 1892; U.S. Patents 510.823, 733.286, 733.287, 733.288, Government Documents, Philadelphia Free Library, Logan Square. See also Ken Butti and John Perlin, *A Golden Thread: 2500 Years of Solar Architecture and Technology* (Palo Alto, Calif.: Cheshire Books, 1980), 102–6.

21. Butti and Perlin, *Golden Thread,* 63–102.

22. "Delaney & Co.'s Plant," in *Tacony Souvenir Program; Boyd's Business Directory of 1900* (Philadelphia, 1900).

23. "The L. Martin Company," in *Tacony Souvenir Program.*

24. "Rose-Tacony Crucible Co.," in *Tacony Souvenir Program.*

25. "6 Fireman Hurt in Blaze," November 12, 1919, "Flames Destroy Tacony Factory," *Philadelphia Bulletin,* May 29, 1926; "Centennial Preserved in Glass," October 15, 1972—all in Gillinder Glass Envelope, *Bulletin* Collection, Urban Archives, Temple University. See also "Gillinder & Sons," in *Tacony Souvenir Program;* and *Philadelphia: Old and New* (Philadelphia, 1895), 82.

26. "France Packing Co.," in *Tacony Souvenir Program.* See also "Right to Quit Jobs Upheld by Court," September 24, 1946; "Firm Gets Right to Sue Strikers," February 17, 1948; *Bulletin* Collection, France Packing envelope, "Control of Stock Is Basis of Suit," *Bulletin,* September 13, 1946; "France Packing Company Sold," November 3, 1965—all in France Packing envelope, *Bulletin* Collection.

27. Recollections of the author, from his boyhood in Tacony.

28. Samuel C. Willits, "History of Lower Dublin Academy," 2:143, Holmesburg Public Library, Philadelphia; interview with Fred Rodgers, grandson of Jane Marsden Dixon, December 14, 1989.

29. Disston Real Estate Sales Book No. 1; interview with Philadelphia trolley historian David Horowitz, July 7, 1988; Holmesburg Library Scrapbook.

Peter E. Costello is the most important political figure in the development of the Far Northeast section of Philadelphia. As a city councilman, he and "Sunny" Jim McNichol formed a coalition to benefit both of them. William D. Disston mingled with both and supported them financially. Disston and McNichol served on the Franklin Institute board of directors. The politics of Tacony evolved around this triumvirate. See "Scrapbook of Peter E. Costello, Public Works Commissioner from December 16, 1903, to June 30, 1904," Peter E. Costello Collection, Atwater Kent Museum; see also *Journal of the Franklin Institute* (Philadelphia), 1913, 339.

30. A book containing copies of the *Disston "Bits"* was kept in the office at Disston & Sons until 1955. When H. K. Porter purchased the firm, there was a general housecleaning of the office. William Rowen, a smither at Disston, took the *Disston "Bits"* book home with him rather than have it discarded. The articles in the "*Bits*" provide a wealth of material about the life of Disston workers and the Tacony community between 1917 and 1920, especially the March 1918 issue, p. 78. A major financial supporter of the Hop, Toad & Frog Trolley Line was Jacob Disston. See "Frankford, Tacony, Holmesburg Minutes Book, 1910–1917," Cox Collection, Atwater Kent Museum.

31. Interview with David Horowitz, July 7, 1988.

32. Interview with Edward Darreff, August 13, 1988.

33. In addition to the atlases cited in Chapter 2, note 42, see *Atlas of Philadelphia,* (Philadelphia, 1920), 23rd and 41st Wards at the Historical Society of Pennsylvania, Philadelphia.

34. Visit by the author to Tacony's St. Vincent's Orphan Home, St. Leo's Church, and Our Lady of Consolation Church in June 1988. Interviews with longtime members of the Tacony community: Harriet Seed of 6606 Torresdale Avenue, and Frank Derenzis, vice president of the United Independent American Club of the City of Philadelphia, 7229 Tulip Street. The charter on the wall of the club indicates that on the founding board of directors in 1896 were Joseph Trailo, 1024 South Ninth Street; G. D. Talone, 624 Carpenter Street; Giovanni DiGiorgio, 927 South Eighth Street; and Nicholas Turchi, 1041 South Ninth Street. The officers came from South Philadelphia, where they lived within a block of one another.

35. John F. Sutherland, "A City of Homes: Philadelphia Slums and Reformers, 1880–1920" (Ph.D. diss., Temple University, 1973), 230; Department of Labor, Negro Division, *Negro Migration During the War* (New York: Arno Press, 1969).

36. Interview with Joseph Rosenwald, April 1, 1989, on his ninety-fourth birthday; interview with Judith Zaslofsky, daughter of Tacony's most trusted and successful lawyer, Hyman Rubin, December 16, 1989.

37. Interview with Bob Bachman, October 13, 1988. This fire led to an investigation of the weak water pressure in the pipes of the Tacony Water Company, which in turn prompted the sale of the water company to the city for more than $900,000. From 1922 on, Tacony's water supply came from the city's Torresdale plant (Holmesburg Scrapbook).

38. *Disston "Bits,"* March 1918, 88.

39. *Tacony,* 8.

40. *Disston "Bits,"* January 1920, 103.

41. Ibid., December 1919, 30–31.

42. Ibid., 30.

43. Ibid., May 1918, 108.

44. Ibid., 109.

45. Ibid., March 1918, 85.

46. Ibid., May and August 1918, 85.

47. Ibid., May and October 1918, 58, 109.

48. Interview with Naomi Vickers, daughter of Billy Seed, March 21, 1989.

49. *Disston "Bits,"* June 1918, 120; January 1919, 106.

50. Interview with Naomi Vickers, January 4, 1990.

51. *Disston "Bits,"* February 1919, 123, 203.

52. Ibid., June 1919, 143, 203.

53. Ibid., May 1919, 180.

54. "Disston History," 1:110–14.

55. Ibid., 1:101–2.

Chapter 5

1. Tamara R. Hareven, *Family Time and Industrial Time: The Relationship Between the Family and Work in a New England Industrial Community* (New York: Cambridge University Press, 1982), 38–39; Chandler, *The Visible Hand,* 260–81.

2. Lizabeth Cohen, *Making a New Deal: Industrial Workers in Chicago, 1919–1939* (New York: Cambridge University Press, 1990), 1–9. Cohen uses "welfare capitalism" to describe these programs, while Alfred Chandler, *The Visible Hand,* uses the term "industrial welfare." The wording has little to do with any difference in what they are describing. I have chosen to use "industrial welfare."

3. *Disston "Bits,"* July 1917, 3.

4. Ibid., August 1917, 1. In comments written to the author, William L. Disston states that nonfamily members were managing the Steel Works, File Works, Jobbing Shop, Saw Works, and General Sales, and that a non-Disston was vice president of finance. Nevertheless, the authority to make the final decision always rested with the Disston family.

5. *Disston "Bits,"* June 1919, 190.

6. "Disston History," 1:112.

7. *Disston "Bits,"* July 1918, 4.

8. Ibid., September 1917, 21; March 1918, 76.

9. Ibid., July, September, and November 1917, 13, 21, 39; January 1918, 60.

10. Ibid., January 1918, 60; July 1919, 5.

11. Ibid., December 1919, 105.
12. Ibid., September and November 1917, 21, 39; February 1918, 60, 72.
13. Ibid., September 1919, 54.
14. Ibid., September 1917, 21; September 1918, 43.
15. Ibid., February 1918, 72.
16. Ibid., May 1919, 168.
17. Interviews with William Rowen, December 7, 1988, and Marguerite Dorsey Farley, September 22, 1988 (Marguerite Dorsey Farley died in 1992).
18. *Disston "Bits,"* January 1918, 61.
19. Ibid., January 1920, 103.
20. Ibid., February 1918, 72; February and April 1919, 123, 154.
21. Ibid., December 1919, 30–31.
22. Ibid., May 1918, 108.
23. Ibid., January 1919, 2.
24. Ibid., February, May, June, and August 1919, 121–22, 180, 201, 30–31.
25. See pictures of children at Disston Recreation Center in 1916, Tacony Manuscript Collection, Tacony Public Library.
26. *Disston "Bits,"* February 1919, 121–22.
27. Interview with Marguerite Dorsey Farley, August 26, 1988; Harry C. Silcox, *The History of Tacony, Holmesburg, and Mayfair: An Intergenerational Study* (Philadelphia: Brighton Press, 1992), 71–77.
28. "Disston History," 1. See pictures of the Disston Recreation Center, the Tacony Public Library, and the Mary Disston School, with dates, found in the Disston picture collection at the Tacony Public Library.
29. *Crucible,* August 1915, 4–6.
30. William Leeds Disston, "Henry Disston and Sons Saw Company History, 1917–1955," in Silcox, *History of Tacony, Holmesburg, and Mayfair,* chap. 4, 48–58.
31. *Philadelphia Inquirer,* March 30, 1954; interview with Marguerite Dorsey Farley, September 22, 1988.
32. *Disston "Bits,"* February 1918, 69.
33. Interview with Marguerite Dorsey Farley, September 22, 1988.
34. Ibid., Marguerite Dorsey Farley remembered the postman's messages as if it were yesterday.
35. *Public Ledger,* August 17, 1918.
36. *Disston "Bits,"* September 1918, 31–32.
37. Ibid., 35.
38. Ibid., May, July, August, and September 1918, 112, 7, 20, 35.
39. Ibid., September 1918, 33.
40. Ibid., 30. The Oxley Post of the American Legion is located at Torresdale Avenue and Rhawn Streets in Tacony. It has as part of its collection the plaque with the following eighteen names listed: Horace W. Ayers, Anthony V. Borucki, Albert W. Buckner, Edward H. Cantz, Nicholas Crispi, John J. Crowe, Benjamin H. Fisher, Marshall B. Lever, Herbert S. Lytton, Harold B. Merz, Leo T. McCabe, Giacono Moscarello, Patrick O'Brien, William D. Oxley, George W. Roberts Jr., Earl W. Schalck, Edward F. Smith, William H. Thompson.
41. Ibid., July 1919, 4.
42. Interview with Bob Bachman, October 23, 1988 (Bachman died in 1990).
43. *Disston "Bits,"* July 1919, 4–5.
44. Ibid.
45. Minute Book, General Factory Committee, September 29, 1919, presently in the possession of the author.

45. Interview with Marguerite Dorsey Farley, September 22, 1988.

46. *Disston "Bits,"* February 1919, 111. In Tacony most men served what might best be described as a primitive M.A.S.H. unit. Tacony was fortunate in that such service had a low casualty rate. The returning Tacony men could proudly boast of having served their country, and proud wives and children echoed their claims. Less fortunate communities, which had infantry National Guard units, lost almost their entire male populations. Death brought to these communities a distaste for war, proving what every reasonable human being knows—that there is no glamour in death. Naturally, there was less patriotism in such communities, only mourning and sadness for the loss of loved ones. Such disparate mortality rates among communities forced the federal government to change the method of assigning troops into the Army during World War II.

The letters home from the servicemen allowed no one on the front line in France to be a coward. Dr. Keiser was in command in France and, before the war and after, was an influential member of the community. Keiser lived until March 28, 1954, when he died in his Tacony home (6933 Tulip Street).

47. Minute Book, General Factory Committee, July 21, 1919, 1.

48. Ibid., 2–3.

49. Ibid., 4.

50. Ibid., 2.

51. William Leeds Disston, "Henry Disston and Sons," 48–58.

52. Ibid.

53. Cohen, *Making a New Deal,* 140–58.

Chapter 6

1. Interviews with Russell McIntyre, June 26, 1990, and William Rowan, December 27, 1990.

2. Fitch, *Steel Workers,* 57–71; interview with Bob Bachman, October 23, 1988.

3. Interview with Thelma Koons, April 24, 1990; *Disston School Messenger,* March and July 1914, May 1915, and May 1919. What the community believed was important for its schools can be observed in these books. Industry needs workers who show up; therefore, schools honored children with perfect attendance by placing them on the cover of school publications. Outstanding grades or schoolwork was not viewed with the same importance by teachers in small industrial communities like Tacony.

4. Milton J. Nadworny, *Scientific Management and the Unions, 1900–1932* (New York: Cambridge University Press, 1955), 1–8; Morgan, "Family Enterprise I," 18. A newspaper reporter who visited the plant in February 1915 described the effects of the company's paternalistic and piecework system on morale through a dialogue with a worker he met afterward on a train to Philadelphia (*Crucible,* December 1915).

> *Worker:* Too bad there is so much of us under the surface, good stuff you can't get hold of by a walk through the shops.
>
> *Reporter:* How's that?
>
> *Worker:* Why, the feelings of the men, their pride in the institution, you cannot see that.". . . I started as an office boy, and now I am a salesman, or they call me one. Really, it isn't difficult to be a salesman for Disston goods; they sell themselves. That's what it means being with a concern with the name. Saves all controversy with the trade. I wonder if the big office men showed you our principal aristocrat? He is just a plain artisan on piecework, but no king has anything on him in Tacony. He quits every day at 12:15 by the clock. He is so skilled and rapid that he works less than six hours continuously all day.

Reporter's comment to himself: A shrewd, optimistic viewpoint that only could have been produced by special and unusual working surroundings. *Crucible,* December 1915.

5. Minute Book General Factory Committee, September 29, 1919, Atwater Kent Museum. William D. Disston recommended a 1,000-point merit system based on predescribed value points for specific parts of the job. Judgments about pay for foremen would be based on the following points: quality of production, 350; quantity of production, 250; method of paying employees; labor turnover, 50; attendance of employees, 100; interest in operating methods, 100; years of service, 50.

6. *Disston "Bits,"* September 1918; Chandler, *The Visible Hand,* 276–77.

7. Black Untitled Minute Book, General Factory Committee, May 7, 1919–May 25, 1920, Disston Account Book Records, Disston Precision files; Disston Journal Nos. 1 and 2, and Disston Private Ledger Nos. 4 and 5, both in Disston Precision files; interview with Roland Woehr, March 27, 1989. Woehr went on to describe how Disston & Sons still had pre–World War I equipment in the 1950s when the plant was purchased by Thomas Mellon Evans. See also Minute Book, General Factory Committee, September 29, 1919.

8. Minute Book, General Factory Committee, December 2, 1919, and March 23, 1920.

9. Interviews with Charles Norbeck, July 28, 1990; Roland Woehr, January 15, 1991; William L. Disston, October 23, 1990; William Rowen, December 27, 1990; Bob Bachman, October 1, 1988; and Thelma Koons, April 24, 1990.

10. Interview with Bob Bachman, October 14, 1988, and Russell McIntyre, June 26, 1990.

11. Interview with Russell McIntyre, June 26, 1990.

12. "Disston History," 1:87; interviews with Roland Woehr, December 27, 1988, August 16, 1989, and June 15, 1991; Thelma Koons, April 24, 1990; and Jack Hansbury, July 17, 1988.

13. Interview with Albert Norbeck, November 21, 1990.

14. Interviews with Russell McIntyre, June 26, 1990; William Rowen, December 30, 1988; Bob Bachman, October 14, 1988. Smither William Rowen and Madison grinder operator Russell McIntyre both kept notebooks with specific company requirements and notes on how to complete certain jobs. McIntyre's book listed what specific companies needed in the product ordered. By using specification sheets, McIntyre was able to provide what customers wanted. Rowen continued Johnny Southwell's practice of placing notations in an order book that gave tensioning instructions and gauge information for specialty saw orders. These were the "quality tools" of the organization.

15. Pocket Black Book, 1924–1927, belonging to Albert Gehr, presently in the possession of the author. Most such books are lost to the historian because workers usually took them with them when they retired. Because the books meant little to the workers' survivors, the books were probably discarded when they died.

16. Early quality practices at Disston resemble those of the W. Edwards Deming Total Quality Movement in Japan (1950s to the present). These pocket black books contain evidence that the workers had more to say about what was quality for a customer than did the bosses in their offices far removed from the shop floors. The system worked then because the communication system between office and shop consisted of "office boys" running messages to the various departments. The system was so slow and cumbersome that the front-line employee *had* to be trusted to make the quality decisions—and they did, and with remarkable success. Only after 1955 and the purchase of the company by H. K. Porter did this system break down. The assembly-line process robbed the worker of the opportunity to make decisions about quality and led to breakdowns in workmanship—and

the Disston product is no longer considered anything special by the public. See W. Edwards Deming, *Out of the Crisis* (Cambridge: Cambridge University Press, 1982).

17. Interviews with Fred Rodgers, November 14, 1988; Bob Bachman, October 13, 1988; and Roland Woehr, August 16, 1989.

18. Interviews with Bob Bachman, October 13, 1988, and Roland Woehr, August 16, 1989.

19. Interview with Bob Bachman, October 13, 1988, and William L. Disston, April 14, 1991.

20. "Hammering and Adjusting Tension of Log Saws in the Filing Room," *Australian Timber Journal* 21 (May 1955), 269–78; interviews with Bob Bachman, October 13, 1988, and Russell McIntyre, June 26, 1990.

21. Interview with William Rowen, January 23, 1991.

22. Interview with John Hansbury, August 24, 1988, and Roland Woehr, December 27, 1988.

23. Interview with Fred Rodgers, November 14, 1988.

24. "Disston History," 1:88–89; interviews with Fred and Arnold Rodgers, December 27, 1988. Samuel Needham told them stories of when he worked with Jonathan Marsden in the Disston Steel plant.

25. Interviews with Ernest and Arnold Rodgers, December 28, 1988.

26. Interviews with William Disston, October 23, 1990; Russell McIntyre, June 26, 1990; Alpheus McCloskey, July 6, 1989; and John Hansbury, August 24, 1988. See also in John Larsen and Ronnie Fader, "Ricky Donahue," Jerry Newman and Harry Silcox, eds., *A Living History of Tacony, Holmesburg, and Mayfair: An Intergenerational Study* (Philadelphia: Printer's Devils, Lincoln High School, 1990), 102–6; and Shane Warner and Susan Lepone, "William H. Rowen," 118–24.

27. Interview with William Rowan, January 23, 1991.

28. Philip Scranton, *Figured Tapestry: Production, Markets, and Power in Philadelphia Textiles, 1885–1945* (New York: Cambridge University Press, 1989), 500–507.

29. Interviews with William Rowen, December 30, 1988; Russell McIntyre, June 26, 1990; Bob Bachman, October 14, 1988.

30. Minute Book, General Factory Committee, September 29, 1919.

31. Ibid., September 29 and October 13, 1919.

32. Ibid., November 17, 1919.

33. Ibid., September 29 and November 17 and 26, 1919.

34. Ibid., September 29, 1919, and October 13 and May 14, 1920.

35. Ibid., September 29 and November 17, 1919; February 10, 1920.

36. Ibid., September 29 and October 13, 1919.

37. Ibid.

38. Ibid., September 29, 1919.

39. Ibid.; interview with Catherine Seed, April 24, 1990, and Thelma Koons, April 24, 1990.

40. Ibid., September 29, 1919.

41. Ibid., September 29 and October 13, 1919.

42. Harry C. Silcox, "Henry Disston's Model Industrial Community: Nineteenth-Century Paternalism in Tacony, Philadelphia," *Pennsylvania Magazine of History and Biography* 114 (October 1990), 483–515.

43. For such advertisements, see the *Crucible,* February 1918–August 1925; *Philadelphia Made Hardware* 3–4 (1911–1912); 21–22 (October 1920–1921); 27–28 (1923–1924); *Disston Lumberman's Handbook* (Philadelphia, 1923).

44. *Philadelphia Made Hardware* 21–22 (October 1923–September 1924), advertisement, 14; "Disston History," 1:10.

45. *Philadelphia Made Hardware* 27–28 (October 1923–September 1924).

46. Ibid. 3–4 (October 1911–September 1912); 27–28 (October 1920–1921). See esp. "Hardware Men See Philadelphia Sites on Way to Atlantic City," ibid., 3–4 (October 1911–September 1912).

47. Interviews with William Rowen, December 27, 1990; Russell McIntyre, June 26, 1990; William L. Disston, October 23, 1990; Mark Ward, June 19, 1990; Bob Bachman, October 13, 1990.

48. Minute Book, General Factory Committee, October 13, 1919.

49. Ibid., March 23 and May 14, 1920.

50. Ibid., September 29, 1919.

51. Ibid., October 13, 1919, and March 23, 1920.

52. Ibid., February 10, 1920.

53. Ibid., January 13 and March 23, 1920.

54. Ibid., February 10, 1920.

55. Ibid., November 26 and December 11, 1919; April 27, 1920.

56. *Disston "Bits,"* April 1920, 54.

57. Minute Book, General Factory Committee, September 29, October 13 and 27, and November 17, 1919.

58. Ibid., September 29 and October 13, 1919; February 10, 1920.

59. *Disston "Bits,"* April 1920, 154.

60. Nadworny, *Scientific Management,* 1–30 (see Chapter 6, note 4).

61. Marvin R. Weisbord, *Productive Workplaces: Organizing and Managing for Dignity, Meaning, and Community* (San Francisco: Jossey Bass, 1989). Weisbord argues that there were two Frederick W. Taylors. The man was viewed by those who knew him "with admiration and affection running on into reverence and worship," but the rank–and–file workers under his system denounced Taylor as "exploitive of labor." The management style of Henry Disston & Sons does not fit Weisbord's generalizations about how management was practiced in America during this period.

62. Interviews with William Rowen, June 30, 1989, and Bob Bachman, October 23, 1988.

63. *Disston Lumberman's Handbook,* 9.

64. Scranton and Licht, *Work Sights,* 178–90; Minute Book, General Factory Committee, December 2, 1919, and March 23, 1920.

65. W. L. Disston, "History of Henry Disston & Sons," 48–58.

66. *Proposed Site for the Sesqui-Centennial, 1926, Upper Roosevelt Boulevard, Tacony Manufacturers' Association* (Philadelphia: New Era Printers, 1921).

67. Holmesburg Public Library Scrapbook, 1922.

68. *Philadelphia Inquirer* article (n.d.) in ibid.

69. *Bulletin,* November 26, 1919, "Man Management Pays at Huge Saw Works in Tacony," in Disston Saw envelope, *Bulletin* Collection, Urban Archives, Temple University.

Chapter 7

1. Cohen, *Making a New Deal,* 129–58.
2. Scranton and Licht, *Work Sights,* 181.
3. Cohen, *Making a New Deal,* 170–79, 189–365.
4. Ibid., 351–54; interview with Bob Bachman, November 26, 1988.

5. Interviews with William L. Disston, October 23, 1990, and William Rowen, March 21, 1989.

6. Interviews with Edward Darreff, July 24, 1988, and William Rowen, March 21, 1989.

7. Interview with William Rowen, March 21, 1989.

8. Interview with John Hansbury, August 24, 1988.

9. Silcox, *History of Tacony, Holmesburg, and Mayfair;* see the interview with George Gross, in ibid., 103–5.

10. Interview with William Hillerman, July 24, 1988.

11. W. L. Disston, "History of Henry Disston & Sons," 58–68.

12. Holmesburg Public Library Scrapbook, 1935; Peter F. Costello Scrapbooks, "Public Works Commission from December 16, 1903, to June 30, 1904," Costello Collection, Atwater Kent Museum.

13. Henry Disston & Sons, *Tool Manual for School Shops* (Philadelphia, 1927).

14. W. L. Disston, "History of Henry Disston & Sons," 58–68.

15. American Trading Co. to N.Y./0-5003-K Japan, July 16, 1934; C. V. Nicholson Export Department to Norman Bye, "Spare Parts for Beet Knife Machine Sent to Belgium," March 31, 1947, in Disston Precision files.

16. Howson & Howson to Henry Disston & Sons, January 2 and 19 and June 10, 1935; envelope containing original patent request by Sam Freas and Norman Bye, in Disston Precision files. The patent was in the name of Bye and Freas, but the profits went to the company because the beet knife was invented on company time.

17. Norman Bye to Mrs. D. Jenkins, "Beet Knife Sharpening Machine," May 7, 1935; Howson & Howson to Samuel T. Freas et al., January 5, 1935, and June 10, 1935; Disston Beet Knife File Parts, N. C. Bye's copy, "Not to be Removed from his Desk," 1947, in Disston Precision files.

18. "Disston Automatic Beet Knife Filing Machine, Beet Sugar Industries Accessories," sales pamphlet in Disston Precision files.

19. Norman Bye to Samuel Freas, May 3, 1935; Howson & Howson to Henry Disston & Sons (Attention Mr. Norman Bye), "Re: German Application," November 29, 1939; Norman Bye to Howson & Howson, December 14, 1939—all in Disston Precision files. Also interview with William L. Disston, January 12, 1991. When the Disston company was sold by the family to H. K. Porter, William L. Disston spoke to Bye about buying out the sugar beet knife business because it had been so profitable to the company. However, the new German product then coming on the market was superior, so Bye recommended against it.

20. Norman Bye, "Report of Visit to Sugar Factories," October 1957; C. H. Slocke to Norman Bye, August 10, 1938; D. W. Jenkins to Norman Bye, "Holly Sugar Inquiry on Knife Grinding Machine," August 17, 1937; R. T. Nalle to Norman Bye, "Improvement Beet Knife Fillers—J. M. Zimmerman, Holly Sugar Co.," January 7, 1938; Norman Bye to Great Western Sugar Co., February 10, 1937—all in Disston Precision files.

21. W. L. Disston, "History of Henry Disston & Sons," 48–58; interview with Bob Bachman, November 22, 1988.

22. W. L. Disston, "History of Henry Disston & Sons," 48–58; interview with William Rowen, March 21, 1989.

23. W. L. Disston, "History of Henry Disston & Sons," 51–58; interview with Roland Woehr, December 29, 1988.

24. W. L. Disston, "History of Henry Disston & Sons," 48–58; interview with Bob Bachman, November 22, 1988.

25. Interviews with Alpheus P. McCloskey, July 6, 1989, and William L. Disston, October 23, 1990.

26. Jeff Rodengen, *Iron Fist: The Lives of Carl Kiekhaefer* (Fort Lauderdale, Fla.: Write Stuff Syndicated, 1991), 91–99.

27. Interview with William L. Disston, January 31, 1992. Disston insists that the Disston company was not trying to get any other company interested in manufacturing the chainsaw engine. The first he knew of Kiekhaefer's plan to dump Disston was when he read Rodengen's book.

28. Rodengen, *Iron Fist,* 94–95.

29. Norman Bye to Jal E. May of Guildford, Australia, April 12, 1961, in Disston Precision files; interview with William L. Disston, October 23, 1990.

30. Rogenden, *Iron Fist,* 96–99.

31. Interview with Roland Woehr, August 16, 1989; interview with Thelma Koons, June 16, 1990, and Alpheus McCloskey, July 6, 1989. William Leeds Disston, in a note to the author in June 1990, disputes some of the statements made by the workers. He said the chain saw "had the greatest potential for the Co. since the handsaw. Those interviewed on this subject were not informed on the chain saw situation in the late 40s and 50s." He further stated that the "chain saw business was very strong until 1955 when H. K. Porter closed the division." These remarks do not answer the many concerns of the employees who were directly working in that division. Russell McIntyre, union secretary, stated that rumors persisted around the plant that $30,000 was needed for research to develop a practical, lightweight, one-man chain saw. The women of the family were reported to be opposed to any such investment. Whatever the reason, there is little doubt that the lack of new products hindered Disston's financial success after World War II.

32. W. L. Disston, "History of Henry Disston & Sons," 55–58; Norman Bye to Jal E. May of Guildford, Australia, April 12, 1961, in Disston Precision files.

33. Norman Bye to Sam Freas, January 3, 1935, in Disston Precision files.

34. *Bulletin,* May 15, 20, and 23, 1941; W. L. Disston, "History of Henry Disston & Sons," 57–58.

35. S. Horace Disston to John Doe, Blankville, N.Y., July 23, 1941 (in scrapbook entitled "Record of Activities Dedicating New Disston Power Plant and the Philadelphia U.S.A. Armor Plate Plant, June 16, 1941," Atwater Kent Museum [hereafter Record of Dedication Activities]).

36. *Bulletin,* December 2, 1939.

37. *Philadelphia Record,* August 12, 1940.

38. *Bulletin,* February 3, 1937.

39. *Time,* May 27, 1940; *Life,* July 7, 1941; interview with Dick Nalle Jr., July 26, 1990. Young Dick Nalle Jr. worked in the Armor Plate building in the summer of 1941.

40. *Philadelphia Evening Ledger,* June 16, 1941.

41. Record of Dedication Activities, 2.

42. Ibid., 10.

43. *Life,* July 7, 1941.

44. H. P. Aikman to S. Horace Disston, August 6, 1941; Joseph G. Terhorst, Mill Supply buyer at Terre Haute Heavy Hardware Co., Indiana, to S. Horace Disston, August 5, 1941. Both letters are in Record of Dedication Activities.

45. Edmund Orgill, Orgill Brothers & Co., Memphis, Tennessee, to S. Horace Disston, July 31, 1941, in Record of Dedication Activities; W. F. Kennedy, Ott-Heiskell Co., Wheeling, W.Va., to S. Horace Disston, July 31, 1941, in Record of Dedication Activities. The Record contains 114 pictures from newspapers throughout the United States and Australia,

including the picture of Disston men taken at the dedication. The quotation appears under fifty of these pictures.

46. *Bulletin,* January 22 and February 27, 1941.

47. Interview with William L. Disston, March 8, 1991; W. L. Disston, "History of Henry Disston & Sons," 48–58. Disston discusses Bye's role in the making of the tank steel plate in great detail.

48. Army and Navy "E" Production Award Ceremonies, December 1, 1942, a scrapbook that contains eighteen newspaper clippings about the event (Atwater Kent Museum).

49. *Northeast News,* December 10, 1942, in ibid.

50. Interview with Catherine Seed, April 24, 1989.

51. Interview with Lawrence "Ricky" Donohue, July 5, 1989; *Nesbitt's Newsbits,* April–May 1967. See also three bound volumes of *Nesbitt's Newsbits:* May 1951 to April 1952, June 1952 to May 1953, and June 1953 to May 1954, in the possession of Lawrence Donohue.

52. I was born on December 1, 1933, and raised in Tacony and remember these kinds of events from my boyhood. My address at the time was 6606 Torresdale Avenue, next door to the Seed family and across the street from Dr. Valentine.

53. *The Disston School Messenger,* March and July 1914, May 1915, and May 1919. Of the teachers listed as faculty members, more than two-thirds lived in Tacony. Also, William Lehr, Edna Whittaker, Emily Wolfenden, and Marian Chambers Streit later taught me in the 1940s at Disston.

54. *Bulletin,* October 14, 1936.

55. Interviews with Bob Bachman, October 13, 1988, and Edward Darreff, August 17, 1988.

56. Recollections of the author.

57. *Philadelphia Inquirer,* April 11, 1943, 1–4.

58. *Bulletin,* February 25, 1943.

59. Recollections of the author.

Chapter 8

1. Interview with Henry Disston, September 24, 1991. He and Morris Disston, son of William Leeds Disston, went to the Disston plant to look over the 1920s records and found that, in their opinion, the dividends paid the family were excessive.

2. Cohen, *Making a New Deal,* 3, 353.

3. Ibid., 5, 252, 257–58, 362.

4. Interview with Edward Darreff, July 24, 1988, and Marguerite Dorsey Farley, September 25, 1988.

5. Holmesburg Public Library Scrapbook, 1922; "Public Works Commission from December 19, 1903, to June 30, 1904," in scrapbooks of Peter E. Costello.

6. *Inquirer,* n.d., Holmesburg Public Library Scrapbook.

7. Interviews with Bob Bachman, October 13, 1988, and William Rowen, June 30, 1989.

8. *Bulletin,* March 20, 1933; interview with William Rowen, June 30, 1989; interview with Thelma Koons, June 16, 1990. Thelma, a worker in the main office at Disston from 1924 to after the company was sold, remembers Dick Nalle as a "handsome man idealized by the women of the office" who had a "wonderful way with the men." He did not work every day, which was the case with William D. and Jacob Jr., but appeared when help was "needed to settle an issue diplomatically with the men." S. Horace Disston was feared by

all workers because of his sharp manner. He worked every day and was the main force in running the business of the firm.

9. *Inquirer,* June 17, 1935; *Bulletin,* June 16, 1935, and February 3, 1935; interview with William Rowen, January 18, 1991.

10. David Montgomery, *Fall of the House of Labor* (New York: Cambridge University Press, 1987), 100–150.

11. Arthur F. McClure, *The Truman Administration and the Problems of Postwar Labor, 1945–1948* (Cranbury, N.J.: Fairleigh Dickinson University Press, 1969), 213–35.

12. Montgomery, *Fall of the House of Labor,* 105–20.

13. *Inquirer,* March 15, 1938.

14. Interview with William L. Disston, October 23, 1990.

15. *Bulletin,* July 13, 1935.

16. *Time,* May 27, 1940, 83–84.

17. U.S. Department of Labor, U.S. Conciliation Service, File 199/245, no. 7374-9.

18. U.S. Department of Labor, U.S. Conciliation Service, Progress Report, March 4, 1939.

19. U.S. Department of Labor, U.S. Conciliation Service, File 199/4649, no. 1057-5A.

20. Interview with William Rowen, December 28, 1988.

21. "100th Anniversary Portfolio," 1–4, Atwater Kent Museum. This portfolio contains samples of all letters, printed pieces, etc., in connection with Disston's open house celebration, May 22–24, 1940.

22. Interview with Roland Woehr, December 27, 1988, and Russell McIntyre, June 26, 1990. McIntyre began working at Disston in 1941 and never heard a union man discuss the bathroom incident. He never knew it happened, even though there is more than sufficient evidence from others that the event did take place.

23. U.S. Department of Labor, U.S. Conciliation Service, File 196/2019, no. 3103-5B.

24. Interviews with William Rowen, March 21, 1989; Bob Bachman, November 22, 1988; and Russell McIntyre, June 26, 1990.

25. "To the Employees," April 26, 1940, in "100th Anniversary Portfolio," Atwater Kent Museum.

26. Steel Workers Organizing Committee to United Saw File and Steel Products Workers of America, April 29, 1940, Urban Archives, Temple University.

27. U.S. Department of Labor, U.S. Conciliation Service, File 196/2019, Exhibit B (in the possession of Phil Scranton).

28. Ibid.

29. U.S. Department of Labor, U.S. Conciliation Service, Progress Report, May 18, 1940 (in the possession of Phil Scranton).

30. U.S. Department of Labor, U.S. Conciliation Service, Preliminary Report, May 3, 1940 (in the possession of Phil Scranton).

31. Interview with William Rowen, March 21, 1989, and Bob Bachman, November 22, 1988.

32. Cohen, *Making a New Deal,* 315.

33. *Bulletin,* March 23, 1940; *Time,* May 27, 1940, 84.

34. *Bulletin,* March 29, 1940.

35. Interview with Roland Woehr, January 15, 1991.

36. U.S. Department of Labor, U.S. Conciliation Service, Progress Report, May 21, 1940.

37. *Time,* May 27, 1940, 84.

38. "100th Anniversary Portfolio," 2.

39. The Rev. Frederick Tyson, Tacony Methodist Episcopal Church, to S. Horace Disston, stamped received May 17, 1940, in ibid.

40. S. Horace Disston to Howard C. Mull, Warren Tool Corporation, July 19, 1940, in ibid.

41. Interview with William L. Disston, October 23, 1990, and Russell McIntyre, January 16, 1991.

42. Minutes of the Executive Board of Local 22254, Disston Union Papers, Urban Archives, Temple University.

43. *Bulletin,* February 24, 1943, Disston envelope, Urban Archives, Temple University.

44. "Grindstone Removal and Change," in Local 22254 correspondence, 1945–1953, Disston Union Papers.

45. Ibid.; *Bulletin,* May 15, 20, 22, and 23, 1940; Disston Union Dues Books, Disston Union Papers. Union Secretary Russell McIntyre donated these union records to the Urban Archives, Temple University.

46. *Bulletin,* December 2, 1939; interview with smither foreman William Rowen, February 28, 1991, and union secretary Russell McIntyre, June 26, 1990. Both men agreed that each foreman set a standard for the amount of work to be done each day. As Rowen put it, "You knew if a man was a slacker and you did something about it." McIntyre stated that the number of arguments over what was a correct amount of work made writing production rules for every job necessary if the workers were going to get a fair hearing on low-production accusations. Management agreed with these sentiments because it established a level of production that was a constant, thus making their cost accounting per product more accurate.

47. "Manpower Assignments," in 1954 Contract, Local 22254; "Minutes of Second Meeting to Discuss Industrial Warehouse" and "Establishment of Seniority for Diesel Locomotive Crane," in correspondence, Local 22254, 1945–1953, Disston Union Papers.

48. Interview with Russell McIntyre, June 26, 1990; "Minutes of Second Meeting to Discuss Industrial Warehouse."

49. W. L. Disston, "History of Henry Disston & Sons," 48–58.

50. Interview with Russell McIntyre, June 26, 1990.

51. Interview with William Rowen, June 30, 1989, and Russell McIntyre, union secretary/treasurer 1973–1988, June 26, 1990. Upon joining the union in 1941, McIntyre was warned of "bad blood" between the steel men and the saw men. William L. Disston commented that a lack of unity in the union made negotiations difficult for management. Disston management basically allowed the union to settle issues of seniority. William L. Disston, June 1990 notation on manuscript, in possession of the author.

52. Minutes of the Executive Board, Local 22254, Disston Union Papers.

53. *Time,* May 27, 1940, 83.

54. McClure, *Truman Administration and Postwar Labor,* 162–84.

55. "Agreement, Effective March 1, 1948," 1–36, in Disston Union Papers.

56. Interview with Russell McIntyre, June 28, 1990.

57. Correspondence, Local 22254, 1945–1953. See also "Hopper Grinding Learner," Disston Union Papers.

58. *Bulletin,* March 5, 8, 10, 21, 23, and 24, 1948.

Chapter 9

1. Barry Bluestone and Bennett Harrison, *The Deindustrialization of America: Plant Closings, Community Abandonment, and the Dismantling of Basic Industry* (New York:

Basic Books, 1982), 3–48; Staughton Lynd, "The View from Steel Country," *Democracy*, Summer 1983, 21–33.

2. Interview with Lawrence "Ricky" Donohue, July 5, 1989; *Nesbitt's Newsbits*, April–May 1967. See also three bound volumes of the *Nesbitt's Newsbits*, May 1951 to April 1952, June 1952 to May 1953, and June 1953 to May 1954, in the possession of Lawrence Donohue. My brother Edward Silcox learned the air conditioning trade at Nesbitt working alongside Franklin Mott, later assistant to the secretary of labor in Pennsylvania, Bill Batty, and many other Taconyites.

3. W. L. Disston, "History of Henry Disston & Sons," 48–58.

4. Interview with Roland Woehr, December 27, 1988.

5. Interviews with Bob Bachman, October 13, 1988, and Roland Woehr, December 27, 1988. Auditor's Report of Financial Statements of Henry Disston & Sons Inc., 1948, 1949, 1950, 1951, 1952, 1953, in Disston Precision files.

6. Interview with Henry Disston (Jacob Jr.'s grandson), July 16, 1988, and William Rowen, June 30, 1989.

7. *Bulletin*, May 23, 1950.

8. Interviews with Dick Nalle Jr., July 12, 1990, and William L. Disston, October 23, 1990.

9. Interview with Roland Woehr, August 16, 1989.

10. *Time*, May 27, 1940, 84. In 1940 there were six Disston men at the plant. The *Time* article takes note of the training then being received by William Leeds "Young Bill" Disston, William D. Disston's son. His training was the same as the previous generations' apprenticeship in the shop.

11. Portfolio containing minutes of the May 14, 1952, board of directors meeting, in Disston Precision files. The list of stockholders and shares owned is on a separate sheet in the portfolio. In an interview with William L. Disston, May 13, 1990, he stated that Evans appraised the Disston real estate at $6 million in 1955 when its actual value was closer to $12 million.

12. Interviews with William Rowen, December 27, 1988; Bob Bachman, October 13, 1988; and Roland Woehr, December 27, 1988. No less a figure then Dick Nalle Jr. admitted that the Disston family did not have the people to manage the firm under the demands of the international market in 1953 (interview with Dick Nalle Jr., July 27, 1990).

13. *Bulletin*, August 18, 1950.

14. *Bulletin*, December 22, 1952.

15. *Bulletin*, April 10, 1954.

16. Rodengen, *Iron Fist*, 1991, 121–23.

17. W. L. Disston, "History of Henry Disston & Sons," 48–58.

18. Rodengen, *Iron Fist*, 121–22.

19. Interview with William L. Disston, October 23, 1990. William L. Disston first mentioned this letter to the author in conversation. Rodengen quotes directly from the letter on pp. 123–30 in his book *Iron Fist*.

20. Interview with Thelma Koons, April 24, 1990, and William L. Disston, October 23, 1990. Koons remembers that the talk in the office was that Gebhart was responsible for the sale of the company, something she could never understand because "Mr. Gebhart was such a wonderful man."

21. W. L. Disston, "History of Henry Disston & Sons," 48–58.

22. Interview with William L. Disston, October 23, 1990. Another alternative to the sale of the company was the selling off of individual parts of the plant, which directive Mellon Evans followed after he purchased the firm. The family, however, refused even to think about downsizing the company.

23. Interview with William L. Disston, October 23, 1990. Disston believes strongly that the sale of the company came to pass because of the failure of the chain-saw operation. The bank stopping payment on all dividends to the directors was going to happen. Another strong argument can be made that the failure of the chain-saw operation was symptomatic of more serious organizational flaws in the company. The families' lack of interest in events until the operation ceased, the lack of communication between an extended family ownership, the lack of leadership to see the venture to its completion, the apparently wrongheaded decision to end a project based on a loan repayable on its cancellation—all raise serious questions about the company's ability to function in a crisis. The question then becomes what incident will bring out the flaws in management—the chain-saw failure or some other event.

24. *Bulletin,* January 11, 1971.

25. Interview with William L. Disston, October 23, 1990.

26. *Inquirer,* November 10–14, 1955 (newspaper clippings given to me by Henry Disston, grandson of Jacob Disston).

27. *Bulletin,* October 19 and November 9, 11, and 15, 1955.

28. *Business Week,* November 26, 1955, 118.

29. Interview with William L. Disston, October 23, 1990; interview with Henry Disston, September 18, 1991. Henry's father, Jacob Jr., did not want to discuss the closing of the plant, so it is impossible to give a firsthand account of events from that side of the family's point of view.

30. W. L. Disston, "History of Disston & Sons," 48–58.

31. Interview with Roland Woehr, December 27, 1988. There are at least two versions of what went on within the Disston family at the time of the sale. Henry Disston, grandson of Jacob Jr., states that his grandfather told him he was in his office when representatives from H. K. Porter visited to make an offer for the company. When Jacob refused the offer, he was informed that Samuel Mellon Evans already owned the majority of the company stock and that Jacob was no longer in charge of the company's operation. William L. Disston, son of William D. Disston, claims that Jacob Jr. "helped engineer the sale to HKP at the outset." William's story seems more likely, since he was directly involved in the transactions during the sale, but the conflicting versions of the story give some idea of the conflicts within the Disston family over the company's sale. Interview with Henry Disston (Jacob Jr.'s grandson), July 16, 1988, and notation by William L. Disston, June 1990.

32. *Business Week,* November 26, 1955, 118. See also *Bulletin,* October 8, 1951; June 29, 1954; and October 19, 1955.

33. *Fortune* 52 (September 1955), 114–16; *Business Week,* January 5, 1957, 118; *Time,* May 11, 1959, 100; *Newsweek,* February 15, 1960; *Forbes,* February 1, 1968, 15–16.

34. *Fortune* 52 (September 1955), 114–16; 61 (May 1960), 142–45; *Business Week,* January 5, 1957, 118.

35. *Time,* June 13, 1960, 103–4; "Porter Divisions Boss Themselves," *Business Week,* November 26, 1955, 119.

36. All quotations in the preceding paragraphs are from *Business Week,* "Porter Divisions Boss Themselves," November 26, 1955, 118–21.

37. Interview with Roland Woehr, December 27, 1988. See also "The Tough Boss," *Time,* August 31, 1959, 64; "New Millionaires: How to Make a Fortune," *Time,* December 27, 1954, 64–65; "Never Work Nights," *Newsweek,* January 25, 1954, 76–78; "An Old Business Is Rejuvenated," *Business Week,* October 6, 1956, 118–19.

38. Norman Bye to Ellwood J. Gebhart, August 4, 1942; Elwood J. Gebhart to C. V. Nicholson, "Beet Knives for Belgium," letters dated March 1 and 11, 1947—both letters in Disston Precision files.

39. American Trading Company to Henry Disston & Sons, New York, June 24, 1954; Norman C. Bye to Fred Howard, "Japanese Made Shaving Knife," July 23, 1958—all in Disston Precision files. Also, interview with Roland Woehr, January 15, 1991. Fred Howard of the Seattle office had Walt Lucki do lab tests on the steel used in the Japanese knives. The test showed no special hardness except at the end of the knife. According to Lucki, this area was shown to "be a composite, which is somewhat of a surprise and I am enclosing a sample showing how much steel was left."

40. Memo, Norman C. Bye, "Visit to Seattle Plant on October 1, 1958"; Disston-Seattle Report, "Corres. Referency Hardness of Circ. Plates"; American Trading Company Inc. order forms from Japan, 1954; Folder, "Am. Trading Company, Mchs. for Japan (50 cycle)—1956"—all in Disston Precision files.

41. N. C. Bye to B. P. Franklin, August 27, 1962, in Disston Precision files.

42. Interview with William McIntyre, January 15, 1991; William Rowen, January 15, 1991; and William L. Disston, October 23, 1990.

43. "An Old Business Is Rejuvenated," *Business Week,* October 6, 1956, 119.

44. Interview with William Rowen, June 31, 1989.

45. Interview with Charles Norbeck, July 28, 1990.

46. Job Description, Records of H. K. Porter Company, October 30, 1958, Disston Precision files.

47. Interview with Russell McIntyre, June 26, 1990.

48. "An Old Business Is Rejuvenated," *Business Week,* October 6, 1956, 120–21; interview with Roland Woehr, December 27, 1988.

49. Interview with Bob Bachman, October 13, 1988, and Roland Woehr, December 27, 1988.

50. Interview with William Rowen, December 27, 1988.

51. Bruce Stutz, *Natural Lives, Modern Time: People and Places of the Delaware River* (New York: Crown Publishers, 1992), 181–90.

52. *Bulletin,* November 30, 1956.

53. List of employees' names and addresses, 1972, found in an abandoned desk drawer in a Disston plant storage area. The desk had been used by the secretary for employee health plans and appears to be one of the few accurate records of employees during the 1970s (list in the possession of the author). Figures listed by the Pennsylvania Industrial Directory are:

Date	*No. of Laborers*
1920	2,425
1927	1,622
1935	1,417
1943	3,122
1956	1,822
1962	542

See *Industrial Directory of the Commonwealth of Pennsylvania 1962* (Harrisburg, Pa., 1962).

54. *Bulletin,* March 7, 1956.

55. *Bulletin,* April 11, 1958.

56. Interview with Roland Woehr, December 27, 1988.

57. "Correspondence Reference: H.S.S. Hard Edge (Super-Safe) Hand Hack Saw"; "Mr. Bye's Visit to Toronto Plant on 8-16-54"; "New Manufacturing Building, Toronto," submitted by Norman C. Bye, Chief Engineer; Memo, Norman C. Bye, "Visit to Seattle Plant

on October 31, 1958"; Disston-Seattle Report, "Hardness of Circ. Plates"; "Hammering and Adjusting Tension of Log Circular Saws in the Filing Room." See *The Australian: The Timber Journal* 21 (May 1955), 269–78 (all in the possession of the author).

58. Interview with Russell McIntyre, June 26, 1990.

59. Interview with Charles Norbeck, July 28, 1990.

60. Interview with Russell McIntyre, June 26, 1990. McIntyre related the hard feelings created when former union men accepted H. K. Porter's offer. When they returned from Danville, the union men of Tacony refused to talk to or have anything to do with them. This boycott is still being carried out by former Disston workers today.

61. "Divisions Disston H. K. Porter Co. De Mexico," report; Norman C. Bye to Albert Strasser, February 17, 1961, in Disston Precision files.

62. N. C. Bye to J. C. Rydick, "Hand Saw Manufacture in Mexico," June 13, 1960, in Disston Precision files.

63. "Porter: Plants, Products, Industries Served, and Sales Offices,"n.d. (in the possession of the author).

64. *Bulletin,* November 26, 1962, and January 11, 1971.

65. *Bulletin,* October 13, 1970.

66. Interview with William L. Disston, October 23, 1990.

67. Interview with Roland Woehr, December 27, 1988, and Russell McIntyre, June 26, 1990.

68. Stutz, *Natural Lives,* 189–90. Stutz writes of Evans's second wife that "distraught over his affections for another woman, his second wife went from room to room of their Virginia home shooting herself with a shotgun, . . . proof enough that Evans brought tragedy in his wake." See also Southly, *Sir Thomas Moore,* 42–55.

69. Hay Management Report, issued by Sandvick Corporation, 1977, in Disston Precision Inc. files; interview with Russell McIntyre, June 26, 1990.

70. Interview with Roland Woehr, December 27, 1988.

71. Interview with Disston Precision Inc. industrial worker Mark Ward, June 18, 1990.

Epilogue

1. *Inquirer,* "Northeast Neighbors Section," January 21, 1988.
2. Interview with Ron Turfitt, August 5, 1990.
3. Bluestone and Harrison, *The Deindustrialization of America,* 19–20.

Index

CPSIA information can be obtained
at www.ICGtesting.com
Printed in the USA
BVHW031839111220
595505BV00011B/84

9 780271 030753